ebb and flow

Tides and Life on our Once and Future Planet

ebb and flow

Tides and Life on our Once and Future Planet

TOM KOPPEL

THE DUNDURN GROUP
TORONTO

Editor: Barry Jowett
Copy editor: Andrea Waters
Design: Alison Carr
Printer: Marquis

Library and Archives Canada Cataloguing in Publication

Koppel, Tom
    Ebb and flow : tides and life on our once and future planet / Tom Koppel.

Includes bibliographical references.
ISBN 978-1-55002-726-6

    1. Tides. I. Title.

GC301.2.K66 2007   551.46'4   C2007-903543-4

1   2   3   4   5   11   10   09   08   07

We acknowledge the support of The Canada Council for the Arts and the Ontario Arts Council for our publishing program. We also acknowledge the financial support of the Government of Canada through the Book Publishing Industry Development Program and The Association for the Export of Canadian Books, and the Government of Ontario through the Ontario Book Publishers Tax Credit program, and the Ontario Media Development Corporation.

Care has been taken to trace the ownership of copyright material used in this book. The author and the publisher welcome any information enabling them to rectify any references or credits in subsequent editions.

J. Kirk Howard, President

Printed and bound in Canada.
Printed on recycled paper.

www.dundurn.com

Dundurn Press	Gazelle Book Services Limited	Dundurn Press
3 Church Street, Suite 500	White Cross Mills	2250 Military Road
Toronto, Ontario, Canada	High Town, Lancaster, England	Tonawanda, NY
M5E 1M2	LA1 4XS	U.S.A. 14150

In fond memory of *Lili Marlene, Seawolfe,* and the *Admiral,*
three fine wooden boats that carried me safely through many a tide.

Table of Contents

Acknowledgements

This book is the result of intermittent work, including research and interviews for magazine articles and newspaper travel stories, that has stretched over more than twenty years. During that time, I had assistance from numerous individuals, many of whom are named in the book. Some of them may not even remember me, but I have retained the notes, tapes, or tape transcripts and remain grateful for their contributions. Others, not mentioned in the text, have assisted me in a variety of ways, from obtaining published source materials to recommending other people as contacts or providing references for grant applications. I also want to recognize some of the editors who gave me assignments for key articles dealing with the tides or related nautical topics. Hearty thanks, therefore, to: Rachelle Balinas Smith; Jean Barman; Peggy Vollmer; Annalee Greenberg, Sandra Barr; John Shaw; Con Desplanque; David Mossman; Chip Fletcher; Mark Merrifield; Jim Dempsey; Mary Lou Bevier; Linda Broussard; John Harper; Brian Robertson; Alan Power; Trent Mick; Andrew Scott; Keith Austin; Ian Darragh; Eric Harris; Cathy Collins; Alan Morantz; Robert Cameron; Shinobu Verhagen; and Louie and Bunny Lorentsen.

It has been a financial challenge to devote the necessary time to this book, and especially to cover the costs of extensive travel. I want to thank both the British Columbia Arts Council and the Canada Council for the Arts for their generous grants. I also received important support, mainly in the form of free or heavily subsidized accommodations and transportation, from several individuals and their organizations for the travel writing assignments that enabled me to gather much of my material. For that, I am grateful in particular to: Mary Lou Foley, Nancy Daniels, and Outrigger Hotels; Stu Glauberman and Aloha Airlines; Randy Brooks and the Nova Scotia Department of Tourism, Culture and Heritage; Mary-Ann Hurley-Corbyn and the New Brunswick Department of Tourism and Parks; Stephanie Curran and the Canadian Tourism Commission; Angela Mah and Air Canada; and Jonathan Reap and Tahiti Tourisme.

I have nothing but the highest praise for my agent, Nat Sobel, and his patient efforts on my behalf. Kudos go to my Dundurn editor, Barry Jowett, and copyeditor, Andrea Waters, and designer, Alison Carr, for their careful help with the text. And finally, I want to express gratitude to my wife, Annie Palovcik, both for her loving support and for proofreading and critiquing my writing at every stage.

Prologue

When I was a child, a high point of each summer was the week my family spent visiting Star Island in the Isles of Shoals, which lie well out in the Atlantic and straddle the state line between New Hampshire and Maine. These were wild, rocky, exposed islands, pounded by the open sea, swept by bitter winter winds, and permanent home to only a few hardy lobster fishing families. Only a single tree grew there, and that was tucked into the shelter of the old Oceanic Hotel where we stayed, and which was open during just the summer months.

It was on Star Island that I first became aware of tides, and of how the whole look, smell, and feel of the place could change in a few hours, depending on whether the tide was high or low. If we arrived on the tiny passenger ferry at low tide, we had to carry our luggage from a floating dock and up a steep ramp to a long pier that towered over the boat. Tides in the Isles of Shoals rise and fall as much as four metres (thirteen or fourteen feet) twice a day. The pier *had* to be long to reach out into water deep enough to accommodate the boat. And the dock where the boat tied up had to rise and fall with each tide.

When the tide was high, the shorelines were neat and sharply defined, with grass or dry, wind-scrubbed rocks extending right down to the high tide line. It was easy and safe for kids like me to walk or clamber along the fringe of the sea. The ocean breeze made everything smell clean and fresh. But at low tide, the shoreline became a ragged and treacherous obstacle course of slippery sloping rock faces and boulders, all draped heavily in brown and green seaweed and encrusted with sharp barnacles. In many places this intertidal zone was twenty or thirty metres (sixty-five to one hundred feet) wide. On hot days, all that seaweed baked in the sun for hours, exuding a powerful miasma of salt and organic decay.

Low tide was the most interesting time to explore, and my parents enjoyed it as much as I did. My mother was an avid amateur naturalist who collected wildlife field guides and loved observing and identifying birds, insects, and almost anything that might creep, crawl, swim, or fly. Often joined by other children, we would pick our way down through the slimy upper belt of barnacles and seaweed to flat, rocky areas where warm tide pools teemed with life. Tiny green and yellow crabs skittered about in the clear, shallow water. Small fishes — sticklebacks, I think — darted under rocks when we leaned in close and our shadows fell on them. There were worms and snails and gloppy, jelly-like little blobs of some kind. Mushroom-shaped anemones in colours from green to orange pulsated and swayed gently on their stalks when a wave spilled over into the tide pool.

The surrounding rocks and low cliffs were festooned with distinct bands of growth, each depending on its distance above low tide. High up were scatterings and solid encrustations of barnacles. Much lower down were various seaweeds, many of them with little air bladders that I enjoyed stepping on or crushing against the rock to hear them pop. And in between were thick bands of blue-black mussels, crowding each other tightly like bunches of grapes.

My father was less knowledgeable about plants and animals than my mother, but he knew what he liked to eat. Back home in the city, he sometimes took me to seafood restaurants or oyster bars and introduced me to the rituals of eating raw oysters and clams on the half shell. Low-tide forays with him on Star Island often turned into mussel-gorging sessions. We simply pried them loose from their clusters, used a handy rock to crack them open, and slurped them down like primitive hunter-gatherers, grinning at each other and feeling mighty proud of ourselves. A real male bonding experience.

Years later, as a young adult on my own, I returned to Star Island for a week. At the time, Cornell University was establishing a marine biology laboratory on adjacent Appledore Island, just across the small harbour. But the facilities were not yet complete, so two or three dozen students were lodged at the Oceanic Hotel and used Star Island for most of that summer's field studies. I sometimes listened in on their lectures and tagged along when they went out to work on their transects. These were very basic field exercises, in which each student ran a fixed line from high tide down to low water and painstakingly counted and logged in a notebook all the organisms they found along that line.

Watching the students and pumping them for details about what they were finding, I began to realize just how diverse the intertidal zone really is in a place like New England, with its large tidal range, i.e. the difference between low and high tides. (It gets much larger as you move farther north along the coast of Maine toward the Bay of Fundy, Nova Scotia, which has the world's largest tidal range, an amazing sixteen and a half metres [fifty-four feet] on some days of the year.) There were scores of different plant and animal species living along those transect lines, and their habitat zones were quite sharply delineated. Every few steps along the line, the distribution of species changed. Both plants and animals were highly sensitive to the amount of time they were exposed, when the tide was out, to air, sun, wind, and predators. Some, like the barnacles high up on the rocks, could close themselves up in their shells and simmer contentedly in the sun and air for hours. They only needed to be immersed in water (and to feed by snagging micro-organisms that swept past them) for a few hours at a time. Others, like crabs, had to stay in the water nearly all the time and were found only down in the lower intertidal. Even there, they were usually either patrolling the tide pools or sheltering in the seaweed. This protected them against the sun and predators, such as the voracious seagulls that hovered on the wind, dove past our heads, and enjoyed rich pickings at low tide.

Then I began to travel, mainly as a journalist and often by ship or small boat, with a particular focus on coasts, islands, and other maritime topics. I noticed how different the shorelines looked where tides were minimal, and how civilization had adapted accordingly. I found that tides were generally quite small at tropical islands well out in the Pacific (such as Hawaii, the Cook Islands, and the Galapagos), along the U.S.

Gulf Coast, and in Florida. By contrast, as in New England, the tidal range is large to very large on the shores of Canada's Maritime provinces and on the coast of British Columbia, where I eventually settled and still live. There the range increases as you go north along the sheltered Inside Passage to Alaska, which has some of the largest tides in the world. The middle latitudes on both U.S. coasts have tides in an intermediate range.

At first I suspected that tides increased consistently from the tropics to the higher latitudes. Then I went to Antarctica and found only small tides lapping on stony, ice-strewn beaches full of penguins. And in tropical Panama, there were large tides on the Pacific side but minimal ones on the Caribbean coast. The way tides work was obviously far more complex than I had imagined. With my curiosity piqued, I decided to find out more about them from scientists and other mariners.

I knew, of course, that tides are caused by the gravitational attraction of the Moon, and to a lesser degree of the Sun, tugging on the ocean. Like most people, I had always thought of them mainly as a slow-acting and benign phenomenon; I visualized the rising tide encroaching on children's sand sculptures and forcing vacationers to move their beach blankets. The picture that emerged, though, is much more dramatic and compelling.

In fact, tides are among the most powerful, inexorable, and often destructive forces in nature, eroding and engulfing shorelines and driving vicious ocean currents, whirlpools, and tidal bores that have claimed countless lives. As an agency of creation, they are largely responsible for everything from the origins of life on Earth to how plants and animals adapt to their particular coastal habitats, from the rotational periods of the Earth and Moon to how we build our port facilities and coastal cities.

Learning to understand and forecast the ebb and flow of the oceans was a scientific and technological task that took humankind centuries. But controversial questions affecting the future of our planet and our species remain. What follows, then, is the story of how tides shape our world. It is a personal journey of discovery and a tale of intellectual challenge and triumph. Its cast of characters includes both recognized geniuses and largely unsung scientific heroes. And its final episodes are still being written in the sky and sand.

High Water and History

India, China, and How the Ancient World Learned
about the Moon and Tides

Our coastal car ferry manoeuvred slowly into the tiny island harbour and docked against the stone seawall. Large doors opened on the side of the ship, and deckhands rolled a short steel ramp into place, bridging the gap between the ship and the cobblestone village street. Hefting my backpack, I walked ashore with the other foot passengers, where I was greeted by a cluster of women who shouted that they had private rooms to rent. Haggling with one old woman, I arranged for a week's accommodation. Meanwhile, a few dozen cars were driving off the ship.

This was Korcula, a long, narrow island full of beautiful and fruitful vineyards just off the Dalmatian coast of what was at the time (the early 1970s) part of Yugoslavia and is now the country of Croatia. I had come for a holiday visit to loll in the sun, sample the local wine and food, and swim in the warm Adriatic Sea. The weather was fine and the main village enchanting, with ancient fortified walls, narrow streets, and intimate outdoor cafés shaded by grapevines that crept over wooden arbours.

One thing about the village and harbour struck me right away, a feature very unlike what I had seen as a child on the coast of Maine, or in British Columbia, where I live today. Where the daily ferry docked, there was no long pier reaching out into deep water, and no need for a steep, adjustable ramp leading from the pier to a floating dock that rose and fell with the tides. Because the Adriatic is part of the Mediterranean, which is largely cut off from the open ocean, the tides at Korcula rise and fall only about one-third of a metre (one foot). This tidal range was so small that the ferry had been designed with doors that were level with the village streets. Cars could roll on and off easily, with no adjustment needed to handle tidal extremes.

In fact, the design of the entire harbour reflected these minimal tides. The seaside streets were only about one metre (just over three feet) above sea level. A seawall ran all along the crescent-shaped harbour with iron rings and cleats set into stone and concrete, so that skiffs and small commercial fishing boats could tie up alongside. People simply stepped down from street level into the boats. The near-absence of tides gave the entire place a tidy, almost manicured, look, feel, and smell. There were no rocks covered in barnacles or mussels to be exposed at low tide, no pungent seaweed baking in the sun. Give or take a few hand spans, the sea always lapped at the shore along the same horizontal line.

It occurred to me that all of the Adriatic and Mediterranean must be this way, which was so different from places with large tidal ranges. I knew that Korcula had been a dependency of Venice when it was a great mercantile city-state. And I realized that a city like Venice, with its network of convenient canals, could never be built in a place subject to large tides. In Venice, only a few steps built into the sides of the canals were needed for people to get into or out of the gondolas. But if a city of canals were built on the south coast of Maine, people would sometimes have to climb ladders four metres (thirteen feet) high and covered in slimy seaweed and barnacles. On the Bay of Fundy, between New Brunswick and Nova Scotia, the ladders would have to be fifteen to sixteen metres (more than fifty feet) high. In other words, the cities and coasts of the Mediterranean only look the way they do because of the very small range of ebb and flow. The Baltic Sea, which is also largely closed off from the open ocean, has a similar tidal situation. When I visited St. Petersburg, Russia, years later, I discovered another city of grand canals where the tidal range was small and where the embankments had only a few stone steps set into them for people to get in and out of boats. There are also many places in the world, as we shall see, that are located on the open ocean and yet have minimal tides. There, too, cities such as Fort Lauderdale, Florida, can be built with easily accessible canals.

Alexander the Great and Tides Beyond the Mediterranean World

The very small tidal range on the Mediterranean had considerable influence on how, when, and where people first began to think about tides and what causes them. Because ancient Greece and Rome were situated on that nearly landlocked sea, the understanding of ocean tides was relatively slow to develop in what we think of as the classical world. Tides are not mentioned anywhere in the Bible, for example. Nor do Plato and Aristotle seem to have been aware of them, except as vague reports sent back by travellers who ventured beyond the limits of the Mediterranean. Herodotus, writing in 440 B.C., devoted only a single sentence in his *History* to the subject when, based on reports from Egyptian sources, he described the Red Sea. "In this sea," he wrote, "there is an ebb and flow of the tide every day." But this was hearsay, not first-hand experience. This lack of knowledge would nearly prove disastrous when Greek and Roman military expeditions ventured into tidal oceans, but that's getting ahead of the story.

Coastal peoples living in places with large tidal ranges must have always been aware of tides. They would have to be in order to coordinate certain kinds of fishing to the regular rise and fall of the sea, and to harvest shellfish safely. (Even today, though, people digging clams and other shellfish, or gathering seaweed, are frequently drowned when the tide comes roaring in faster than expected. This happened to almost two dozen immigrant Chinese cockle pickers on the north coast of England in 2004.) Ancient maritime people would have been observant and cautious, roughly gauging the timing of the tides by the passage of the Moon or Sun. Naturally, they also sought explanations for the causes of the tides and considered how to manage life along a tidal coast. Natives of Malaya thought tides were caused by the movements of a giant crab. The Tlingit of coastal Alaska believed that the tides were caused by Raven, their powerful trickster. In the mythology of the Haida on Canada's Queen Charlotte Islands, there was a character called Flood-Tide Woman who could cause the sea to rise simply by raising her skirt. Peasants in Brittany believed it was best to sow their clover at high tide and that the best butter is made just as the tide begins to flood. The natives of New South Wales, Australia, traditionally burned their dead on a flood tide, because an outgoing tide would carry their souls away to distant places. And many cultures have believed that no creature can die except at the ebbing of the tide.

Some people in ancient India attributed the tides to the pulse or breathing of a monstrous sea god. But it was also on the coast of India that people first dealt with tides on a large scale and in an organized and systematic way. And it was Indians who made the earliest realistic guesses as to the causes of tides.

The Gulf of Cambay, part of the Arabian Sea on the northwest coast of India (today's state of Gujarat), has an extremely large range (vertical distance) between low and high tides. Ocean scientists call a tidal range of up to one metre (just over three feet) microtidal. Four metres or greater (thirteen feet or more) is termed macrotidal. The middle-size tides in between are called mesotidal. At the mouths of some of the rivers that flow into the Gulf of Cambay, the range between the largest and smallest tides of the year is about ten metres (some thirty-three feet). In this book, I will take the liberty of calling tides of that size or greater megatidal. Other places with megatides include Canada's Bay of Fundy and Ungava Bay, Britain's Bristol Channel (also known as the Severn Estuary), the north coast of France, the west coast of Korea, and Cook Inlet in Alaska.

The Gulf of Cambay was on the eastern edge of the ancient Harappan (or Indus) civilization. Artifacts have been discovered (such as a distinctive seal and a potsherd portraying a ship) that show the Harappans traded, probably by sea, with the two other earliest known civilizations, Egypt and Mesopotamia. All were connected along the waters of the Arabian Sea, Red Sea, and Persian Gulf. There were a number of large trading ports in the Cambay region, including Lothal, which was situated inland along the Sabarmati River and could only be reached at high tide. The problem for sizable commercial vessels trading with Lothal was what to do when a high megatide ebbed.

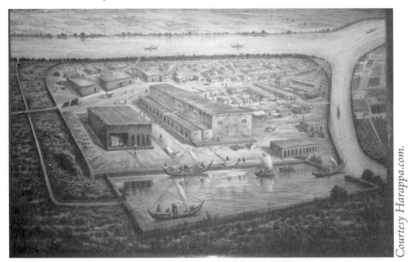

The Lothal tidal dockyard in India, circa 2000 B.C.

It is a difficulty that mariners and naval architects have always faced. Under ideal conditions — this might be within a sheltered bay with a soft sand or mud bottom without obstructions such as boulders or sharp coral heads, when there is little wind and no ocean swell is running — some large ships can be allowed to go dry and sit safely on the bottom through a low tide. This was often done intentionally (it is called careening the ship) to scrape off barnacles and other growth, or to repair damage to the hull. Some boats and ships are even designed with flat, reinforced bottoms, or with well-spaced parallel double keels that allow them to sit upright on the tidal flats of a muddy estuary. But if a ship not designed for such conditions goes aground at the wrong time — such as when strong winds are kicking up sizable waves, or when large swells are rolling into a bay from the open ocean — it can be pounded to pieces either before the tide drops enough to leave it safely high and dry or as the tide comes back in after a low tide. Moreover, the shape of many ships' hulls does not allow them to sit upright when on the bottom. When the tide ebbs, they will roll over onto one side or the other, exposing their decks and any open hatches to pounding waves and possible flooding.

As for ancient mariners on the Arabian Sea, unless their ships were designed with flat bottoms to sit on the mud flats of bays and estuaries between high tides, they could easily be damaged. And even if so designed, they would be difficult or impossible to reach and service at low tide. So the Harappans devised and engineered an elaborate solution for the port of Lothal.

They dug out a tidal dockyard larger than two football fields set end on end and connected it to the tidal estuary by a long, narrow channel. Roughly rectangular in shape, the dockyard was more than 218 metres (716 feet) long on each side. At one end it was more than 34 metres (112 feet) wide, and at the other end more than 38 metres (125 feet) wide. The walls of the embankments, along which the ships could tie up to load and unload, were lined with kiln-fired bricks. Where the channel from the estuary entered the dockyard, a heavy wooden door could be put in place to seal off the entrance. When the dockyard was first built, the channel to the estuary was quite short, but during the centuries that it was in use the river shifted away from the dockyard. The channel had to be extended and became a true canal 2.5 kilometres (over 1.5 miles) long.

The dockyard was large enough to accommodate and service dozens of trading vessels at once. At high tide, they were brought through the channel and into the enclosure. Then the door was closed so that they could float safely during periods of low tide and could be conveniently loaded and unloaded as well. When the tide rose again, ships could leave the dockyard and others could enter. It was a highly efficient solution to the problem of large tidal ranges, and the concept is still used at some ports today. In Seattle, for example, much of the city's commercial fishing fleet and many yachts are not kept on the open waters of Puget Sound, which has large tides. They are brought through a lock and docked in an inland basin called Lake Union, where the water level barely changes.

By modern times, the Lothal dockyard had long since been abandoned and had silted up. But it was excavated in 1958–59 by archaeologists from the Archaeological Survey of India, and the earliest four stages of construction were radiocarbon dated to between 2450 and 1900 B.C. A later stage of construction was dated to between 1900 and 1400 B.C. As those earliest dates show, its construction four millennia ago began only a few centuries after the great pyramids of Egypt were built, making it one of the engineering wonders of the ancient world.

Not surprisingly, the ancient Indians, forced to deal with extreme tides, observed them closely, wrote about them, and speculated on their nature. The earliest clues to such knowledge come from a classical

Sanskrit text, or Veda, the Sama-Veda, a sacred poetic mantra with more than fifteen hundred verses. One of its several versions, or revisions, the Kauthuma Sama-Veda, has traditionally been associated with Gujarat, where the tides are the largest on India's coast. Estimates of when the Sama-Veda was created differ wildly, but it was most likely between 1500 and 1200 B.C. The Sama-Veda makes clear reference to the swelling of the mighty oceans or seas and suggests that this is due mainly to the influence of the Moon. One part of the Sama-Veda speaks of tides as "the gathered flood of oceans."

The connection between tides and the Moon was expressed more clearly in two Hindu epics, the Ramayana and the Mahabharata. The Ramayana, variously dated to between 1000 and 250 B.C., stated that tides are most greatly influenced by the Moon during *parva*. Astronomers today call this *syzygy*, when the Moon is aligned with the Sun, either on the same side of the Earth as the Sun (new moon) or opposite the Sun (full moon). Speaking of the "rise and fall" of the sea during these times of *parva*, the Ramayana refers to "the roaring of the heaving ocean during the fullness of the moon." The slightly later Mahabharata specified that it was during new moon and full moon that the tides were most intense.

Later Sanskrit texts, the Puranas, gave a more refined conception of what was involved when the tides ebbed and flowed. To begin with, they rejected the idea, which was common in ancient times and persisted right up through the Middle Ages, that it might be the flow of waters into and out of the ocean (along rivers, for example) that accounted for the rise and fall of the sea. According to the Visnu Purana, "in all the oceans the water remains at all times the same in quantity and never increases and diminishes." Then the text offered an alternative explanation: "But like the water in a cauldron, which in consequence of its combination with heat expands, so the waters of the oceans swell with the increase of the Moon." As for the timing of this action, "the waters, although really neither more nor less, dilate or contract as the Moon increases or wanes in the light and dark fortnights," that is, around the times of full moon and new moon. In other words, the relative size of the tides followed a two-week cycle, which is essentially correct.

The Matsya Purana was even more precise: "When the Moon is in the east [where it rises], the sea begins to swell. The sea becomes low when the Moon wanes. When it swells, it does so with its own waters (and not with additional waters), and when it subsides, its swelling is lost in its own waters …" Here, again, was a clear rejection of the idea that waters were entering or leaving the ocean from some other source. In summation, the Matsya Purana stated, "On the rising of the Moon, the sea increases as if its waters have really increased. During the bright and dark fortnights, the sea heaves at the rising of the Moon and becomes placid at the wane of it, but the store of water remains the same. The sea rises and falls according to the phases of the Moon." And although the ancient Indians were not implying a solar influence on tides, those phases of the Moon are in fact determined by the Moon's position in the sky relative to the Sun. The Indians had been careful in their observations. The Puranas also provided highly accurate measurements of the tidal range, listing it as 510 angulas (nearly 10 metres, or 32 feet) on *parva* days.

Very large tides in the Arabian Sea caught the attention not only of Indian observers but of the earliest European ones as well. When Alexander ("the Great") of Macedonia marched eastward, he not only defeated the Persians, he carried on right through today's Afghanistan and into northwestern India (now Pakistan). There his army, too long away and disheartened, refused to push on into the heart of India. Alexander relented and decided to head back home. He built a fleet of eighteen hundred boats or small ships, and in 325 B.C. he sailed them south along the Indus River, pillaging and murdering tens of thousands, including women and children, as he went. At Pattala, the Indus divided into two main branches, or distributaries, and formed a delta. Alexander explored both channels to determine which would be the safer route for the bulk of his fleet. According to his plan, part of his army would then return to the west overland, and the other part would go by sea. Although Herodotus apparently knew a bit about tides on the Red Sea over a century earlier, this does not seem to have become known to Alexander's soldiers, who had reached India along an inland route.

Alexander sailed first down the western branch of the Indus, and as his ships approached the mouth, the officers and crews were surprised and horrified to encounter the large tides and fierce tidal currents there. The phenomenon was entirely novel to them. In one Greek account, waves surged into the river from the open sea, causing "great alarm and confusion among the navigators and considerable damage to the Flotilla."

They ran the fleet into the shelter of a canal, but that did not protect them against the tide. The authoritative Greek record, the *Anabasis of Alexander*, went on to relate that, once the fleet was anchored, "the ebb-tide, characteristic of the great sea, followed; as a result their ships were left high and dry." (At this point, the modern translator added a footnote stating, "Tides and tidal bores were unfamiliar to Mediterranean dwellers.")

This left the ships, at least temporarily, in a safe situation, but one that did not last. The *Anabasis* added that "Alexander's men had not known of this [the tides] before, and it was another thing that gave them a severe shock, repeated with still more force when the time passed and the tide came up again and the ships were lifted up." The danger, at that point, depended on how the ships were resting. "Ships which the tide found comfortably settled on the mud were lifted off unharmed, and floated once more without sustaining damage; but those which were caught on a drier bottom and were not on an even keel, as the onrushing tide came in all together, either collided with each other or were dashed on the land and shattered." Damage was extensive, and "Alexander repaired them as best he could."

Little wonder that Alexander's men would be terrified and confused by what they encountered. Sir Alexander Burnes, a British officer who explored the lower Indus in the 1830s, described a tidal regime that was bound to shock soldiers with no relevant experience: "The tides rise in the mouth of the Indus about nine feet [nearly three metres] at full moon and 'flow and ebb' with great violence, particularly near the sea, where they flood and abandon the banks with equal and incredible velocity. It is dangerous to drop the anchor unless at low waters, as the channel is frequently obscured, and the vessel may be left dry [when the tide falls]."

An even more evocative account of tides in India, and of their dangers, was provided in the *Periplus of the Erythraecan [Arabian] Sea* (generally just called the *Periplus*). It was published in A.D. 65 by an anonymous Greek or Egyptian merchant who sailed to the market towns on the Arabian Sea. One important trading emporium was Barygaza (today's Broach), which was well inland along the Narmada, a river that flows into the Gulf of Cambay. The author's ship picked up a local fisherman at the mouth of the river to pilot them safely, and he steered the ship upriver with stops at certain fixed places, "going up with the beginning of the flood [tide], and lying through the ebb at anchorages and in basins. These basins are deeper places in the river."

The *Periplus* gave a vivid description of the tides in that region, and possibly described a tidal bore: "Now the whole country of India has very many rivers, and very great ebb and flow of the tides; increasing at the new moon, and at the full moon for three days, and falling off during the intervening days of the moon. But about Barygaza it is much greater, so that the bottom is suddenly seen, and now part of the dry land are sea, and now it is dry where ships were sailing just before." These large tides were particularly treacherous when they ran opposite to the flow of large rivers. At such times, "the rivers, under the inrush of the flood tide, when the whole force of the sea is directed against them, are driven upwards more strongly against their natural current … [the] entrance and departure of vessels is very dangerous to those who are inexperienced or who come to this market-town for the first time. For the rush of waters at the incoming tide is irresistible, and the anchors cannot hold against it."

The result, frequently, was maritime mayhem. "Large ships are caught up by the force of it, turned broadside on through the speed of the current, and so driven on the shoals and wrecked; and smaller boats are overturned; and those that have been turned aside among the channels by the receding waters at the ebb, are left on their sides, and if not held on an even keel by props, the flood tide comes upon them so suddenly and under the first head of the current they are filled with water."

As with the ancient Indian tomes, the *Periplus* linked these killer tides to the phases of the Moon: "For there is so great force in the rush of the sea at the new moon, especially during the flood tide at night, that if you begin the entrance at the moment when the waters are still, on the instant there is borne to you at the mouth of the river, a noise like the cries of an army heard from afar; and very soon the sea itself comes rushing in over the shoals with a hoarse roar."

During the centuries between Alexander's time and that of the *Periplus*, considerable tidal lore had filtered back to the Mediterranean world, where it was systematically recorded by Greek and Roman geographers. Increasingly, the tides were seen as linked to the passage of the Moon across the sky and through its monthly cycle.

A few years after Alexander's campaign in India, Pytheas of Marseilles, a Greek navigator who was searching for sources of British tin, made a voyage out past Spain and France and circumnavigated Britain, where some of the coasts experience very large tides. The report of Pytheas no longer exists, but he supposedly said that the tide came in as the Moon waxed and went out as it waned. According to the *Natural History* of the Roman encyclopedist and writer Pliny the Elder, Pytheas also claimed that British tides sometimes rose more than 36 metres (120 feet), which would be a great exaggeration over the actual range of over 10 metres (30 to 40 feet or more).

A later and more detailed report on North Atlantic tides came from the Stoic philosopher Poseidonius, who lived from 135 to 51 B.C. He travelled to Cadiz, on the Atlantic coast of Spain just beyond Gibraltar, where he personally observed the tides and gathered information on their timing and long-term patterns from local inhabitants. As with Pytheas, his written reports were eventually lost, but not before the Roman geographer Strabo had quoted from them at length. Poseidonius noted the clear link between the tides and the passage of the Moon in the sky, asserting that "the movement of the ocean is subject to periods like those of the heavenly bodies … behaving in accord with the Moon." He noted daily, monthly, and yearly lunar periods and correlated them to the tides. On a daily basis, "when the Moon rises above the horizon to the extent of a zodiacal sign [30 degrees], the sea begins to swell, and perceptibly invades the land until the Moon is in the meridian," that is, due south of an observer in the northern hemisphere. "But when the heavenly body begins to decline, the sea retreats again, little by little, until the Moon [reaches] a zodiacal sign above her setting; then remains stationary." In other words, there is a period of slack tide. Later in the twenty-four-hour cycle, the sea "invades the land again until the Moon reaches the meridian below the earth." Poseidonius was, therefore, describing a twice-daily or semi-diurnal tide, which is the norm on the North Atlantic. Moving to the monthly pattern, he noted that the overall tidal range became greatest at the time of conjunction, or new moon, when the Sun and Moon are on the same side of the Earth. After that, the tides get smaller until half moon and then become larger again until full moon, which agrees with modern observations.

As for the annual periods, he claimed that the largest tides occurred at the solstices and the smallest at the equinoxes. (This is not a universally valid generalization. A close look at the tide tables for the coast of

British Columbia, for example, shows large tidal ranges on some dates close to both the solstices and equinoxes, but during other months as well.) However, by the time of Pliny the Elder, who lived from A.D. 23 to 79 and apparently had other sources, some informant had noticed an annual pattern that seemed to be the opposite of what Poseidonius had claimed. Pliny wrote that tidal ranges were larger at the equinoxes, and that the largest tides came not exactly at the time of the new or full moon but a few days *after* conjunction or opposition. (This is not universally true either. The situation is, in fact, almost maddeningly complex. On any given day, the size of the tides also depends on the Moon's distance from the Earth and its declination, or degree of elevation in the sky north or south of the Earth's equator.) Pliny was a careful recorder of facts and had great curiosity about the surrounding world. This was his undoing. When Mount Vesuvius erupted, he was located safely distant across the bay, but wanted to observe its effects. He sailed across to the doomed city of Pompeii, got much too close, and died from the ash fall and overheated or poisonous gases.

Whereas at Cadiz the tides followed a pattern of semi-diurnal near equality, Poseidonius pointed out that an earlier astronomer at Babylon, named Seleucus, had noted a very different pattern on the Arabian Sea. This is called the diurnal inequality, where, for certain periods each month, one high or low tide each day is much higher or lower than the other. As we shall see, this is common in many parts of the world. Seleucus thought the regularity or irregularity in the tides depended on where the Moon was in the sky in relation to the signs of the zodiac. That is, when the Moon is in the signs close to the equinoxes, the behaviour of the tides is regular, whereas it is irregular when the Moon is in the signs close to the solstices. According to one modern tidal scientist, this observation of the diurnal inequality and "how it would appear at equinoxes and solstices accords exactly with that observed in all seas adjacent of modern Saudi Arabia, while being barely perceptible at Cadiz." Assuming that the Earth's atmosphere reached right up to the Moon, Seleucus suggested that the Moon caused the tides by alternately pressing down on the atmosphere and resisting the atmosphere's pressure. This pressure or resistance would, in turn, generate winds that influenced the ocean.

Julius Caesar Thwarted by Large Tides

Although bits and pieces of accurate information about the tides had been published by the time of the Roman Empire, people who lived on the Mediterranean still had very little practical experience with them. This nearly led to disaster when Julius Caesar invaded Britain in 55 B.C. At what is today's port of Boulogne, on the French coast, he assembled a fleet of eighty ships to transport two Roman legions, or about ten thousand men, under his personal command. These were to be followed a few days later by eighteen heavier "ships of burden" that carried the horses for his cavalry. After an overnight crossing of the English Channel, Caesar's squadron arrived off the towering cliffs of Dover, where, forewarned, thousands of native Britons were assembled both high on the cliffs and at their base to defend the beach. The official Roman account later reported that "the sea was confined by mountains so close to it that a dart could be thrown from their summit upon the shore." Rightly considering this a crazy place to disembark, "[Caesar] remained at anchor till the ninth hour, for the other ships to arrive there." Then, when the winds and tides were favourable, Caesar moved his fleet a short distance up the coast to the northeast and stationed his forces off "an open and level shore." When the Roman soldiers landed in the surf, "oppressed with a large and heavy weight of armor," the Britons mounted a vigorous defence and enjoyed the advantage of attacking from dry land. Eventually, though, Roman skill and discipline won out, the legionaries stormed ashore, and the "barbarians" sent ambassadors to Caesar to offer him hostages and negotiate a peace, which was established after four days.

Meanwhile, the eighteen ships carrying the Roman cavalry had set out for Britain, but "so great a storm suddenly arose that none of them could maintain their course at sea; and some were taken back to the same port from which they had started." Others anchored off the British shore, where the smaller ships that had carried the troops were still sitting on the beach. The combination of stormy seas and unexpected tides was devastating. As the Roman account tells it, "It happened that night to be full moon, which usually occasions very high tides in that ocean." But this was written with the advantage of hindsight. "That circumstance," the report continues, "was unknown to our men. Thus, at the same time, the tide began to fill the ships of war which Caesar had provided to convey over his army, and which he had drawn up on the strand; and the storm began to dash the ships of burden which were riding at anchor against each other; nor was any means afforded our men of either managing them or of rendering any service." (This is hardly surprising, with such a high tide and waves slashing the exposed beach.)

Twelve ships were totally lost and many others damaged. This left Caesar without the cavalry he had expected to play a major role in his invasion. His forces were also immobilized by the need to repair the remaining ships, which left them at the mercy of the Britons, who saw their chance, broke off the peace negotiations, and renewed their attacks. As the official record described it, "A great many ships having been wrecked, inasmuch as the rest, having lost their cables, anchors and other tackling, were unfit for sailing, a great confusion … arose throughout the army." Caesar ordered a withdrawal and managed to get his forces back to the coast of Gaul. The following summer, he invaded Britain again. Once again, he encountered a terrible storm, and lost many of his ships on the same beach at Kent. After a couple of months of skirmishing in southern England, he withdrew once more. It was nearly another century before the Romans returned to Britain and launched their long-term conquest.

China's Tidal Bore and the Venerable Bede

At the opposite end of the Eurasian landmass, the Chinese were also gradually learning about the tides. Some areas of eastern Asia's coasts and estuaries have very large tidal ranges. In pre-modern China there were several explanations for the tides. One was that water was considered to be the Earth's blood, and the tides were the beating of its pulse. Another idea was that tides were caused by the breathing of the Earth itself. A very different concept was offered in the fourth century A.D. by a writer named Ko Hung, who suggested that the difference in tidal ranges between spring tides and neap tides was because the sky itself moved eastward and then westward once each month. And he claimed that summer tides were higher than winter tides because in summer the sky was much farther away, which meant that the female or negative principle in nature was weak, while the male or positive principle was strong. George

Howard Darwin, the son of Charles Darwin and one of Victorian England's leading tidal scientists, reported in some detail on tides in China and commented on Ko Hung's argument. Darwin confirmed the existence of a strong diurnal inequality on the Chinese coast, "such that in summer the tide rises higher in the daytime than in the night, whilst the converse is true in winter. I suggest that this fact affords the justification for the statement that the summer tides [in China] are great."

The sophistication of ancient China's tidal knowledge was largely due to the phenomenon known as a tidal bore. This occurs on certain rivers in many parts of the world, but only at times when the river's outward flow and the opposing incoming flood tide create specific conditions. Where a river estuary has just the right funnel-like shape, depth, and slope, a very large flood tide will fight and eventually win out in a battle against the powerful downriver flow of fresh water. Just as ocean waves build into tall breakers when the swell reaches the shallows off an open beach, on the lower reaches of a river the swiftly moving tide is slowed by friction as it encounters the shoaling river bottom and builds into a large breaking wave. This can surge upriver as a high and very steep wave front that advances at great speed.

In Canada, there used to be a powerful tidal bore on the Petitcodiac River in New Brunswick, where it flows into the Bay of Fundy. (When a modern causeway for cars was built, it altered the river's flow.) In Europe there are prominent bores on the Seine, Orne, and Gironde rivers in France and on the Severn, Wye, and Trent in England. On the Amazon, in Brazil, the bore is so dramatic that it has a name, the Pororoca. As one modern study of tides describes it, "The Pororoca is often so great as to make the river impassable. From the banks it looks like a mile-long waterfall, up to sixteen feet [nearly five metres] high and travelling upriver with a speed of twelve knots. Its roar can be heard almost fifteen miles [twenty-four kilometres] away."

In China, the largest and most powerful tidal bore occurs on the Qiantang (Qian River) below the large city of Hangzhou. Known as the Dragon or Black Dragon, it is one of that region's prime tourist attractions and is watched by thousands of people each year, especially at an annual festival held in the small town of Yanguan each September. A tidal range of almost six metres (about nineteen feet) in the estuary generates a wave that is well over two storeys high as it races upriver.

In his popular science book *The Tides*, published in 1898, George Darwin described how five or six thousand people assembled on the bank of the Qiantang and threw in offerings to appease "the god of the waters." Then they waited and watched. "This was the occasion of one of the highest bores at spring tide, and the rebound of the bore from the sea-wall, and the sudden heaping up of the waters as the flood conformed to the narrow mouth of the river, here barely a mile [over 1.6 kilometres] in width at low water, was a magnificent spectacle," Darwin wrote. "A series of breakers were formed on the back of the advancing flood, which for over five minutes were not less than twenty-five feet [7.6 metres] above the level of the river in front of the bore."

Darwin reported that for centuries the Chinese were in awe of the bore, which they believed was caused by the spirit of a general who had been assassinated by a jealous emperor and his body thrown into the river. There "his spirit conceived the idea of revenging itself by bringing the tide in from the ocean in such force as to overwhelm the city of [Hangzhou], then the magnificent capital of the empire." This led to the flooding of large portions of the region. The emperor tried to propitiate the spirit by burning paper and making offerings of food along the seawall. "This, however, did not have the desired effect, as the high tide came in as before; and it was at last determined to erect a pagoda" to pacify the sea, which was positioned according to strict Chinese rules to guarantee good feng shui. "After it was built the flood tide, though it still continued to come in the shape of a bore, did not flood the country as before."

In September 2002, however, a typhoon in the South China Sea generated a storm surge and much larger than normal flood tides. Thousands of visitors had gathered atop the seawall, which was supposed to be at a safe height for watching the bore. But it swept in with such violence that the churning wave came crashing right over the seawall, forcing the tourists to run for their lives. A dramatic photo of them fleeing in terror made the front pages of newspapers around the world. Historically, it has drowned countless unwary people.

The bore on the Qiantang was so spectacular that in 1056 (some decades after the construction of the pagoda aimed at pacifying it) a pavilion was also built at Yanguan so that people could wait for the bore and watch it in comfort. By that time, the Chinese had noticed the link between the timing of high tides and the movement of the Moon, which enabled them to produce the world's first tide prediction table. (This was two centuries before the earliest European tide table for the Thames at London Bridge.) The inscription was a series of simple lists giving the day, time of day (in two-hour increments, with each twenty-four-hour day divided into the "Twelve Earthly Branches") when the bore would arrive, and relative size of the predicted high tide. It forecast the time of the bore for each day in the lunar month and for each season of the year. A twentieth-century scientific study compared the table with the actual times of the tidal bore today. The sets of times agreed quite well for the highest (perigean) high tides of the year, but not for the lower (apogean) ones. The predictions, which of course were for bores occurring weeks, months, and years into the future, were off by as much as two hours. Still, as the study noted, "such deficiencies are inherent in such a simple tide-prediction table, but evidently the predictions were good enough for their purpose — fairly accurate for the most spectacular bores, especially for leisurely tourists carrying no watches."

By the time the Chinese were building their pavilion for observing tidal bores, knowledge of tides had advanced considerably in Europe and the Middle East. In England, the Venerable Bede, a monk in North-umbria (modern-day Northumberland), where the North Sea coast experiences large tides, was a careful and lifelong astronomical observer. In part, this was because he sought to determine the date of Easter by watching the Moon. In A.D. 703 he wrote, "The most admirable thing of all is this union of the ocean with the orbit of the Moon. At every rising and every setting of the Moon the sea violently covers the coast far and wide, sending forth its surge, which the Greeks call *reuma*; and once this same surge has been drawn back it lays the beaches bare.... As the Moon passes by without delay, the sea recedes.... It is as though it is unwittingly drawn up by some breathings of the Moon, and then returns to its normal level when this same influence ceases."

Around A.D. 730, Bede noted that high tide came about forty-eight minutes later each day (this is called the daily retardation), reflecting the later passage of the Moon with each successive day of the month. This was quite close to the modern figure of around fifty minutes, and not bad for a time when accurate clocks were rare. He also noticed that strong winds blowing onto the shore, or away from it, could affect both the height and timing of the tides, for example by causing significant storm surges with much higher than usual high tides.

Most significantly, he observed that the tide would rise on one part of the British coast at the same time that it fell elsewhere. "We know, we who live on the many sided shore of the British Sea, that when this sea begins to flow, in the same hour another will begin to ebb." This was a real conceptual breakthrough, because it meant that the sea was not changing its volume by boiling over, as in a cauldron, or by expanding and contracting as the Moon passed overhead. Rather, tides had to involve, in some way, the movement of water from one part of the ocean to another. Nor could there be a simple and consistent correlation between the passage of the Moon across the meridian and the time of high tide in all locations, as many of the Greek and Roman writers had assumed from observations on the Arabian Sea, in Cadiz, and elsewhere. The time of high tide varied with each portion of the British coast, and at least in some areas it did so in a progressive way. Bede observed, for example, that the time of high tide progressed from north to south along the Northumbrian shore. As he put it, "On one and the same shoreline those who live to the North of me will see every sea tide both begin and end much earlier than I do, whilst indeed those to the South will see it much later." Yet he recognized that for any given place, high water always seemed to arrive when the Moon was at the same position in the sky, say a certain number of hours before or after it passed the meridian. (For each port along a coast, however, this would be different.)

Bede also recognized the great complexity in the pattern of tidal flows, picturing them as streams that he thought flowed southward on both sides of Britain and met along the southern coast. "The Isle of Wight," he wrote, "lies opposite the boundary between the South Saxons and the Gewissae, and is separated from it by three miles of sea, known as the Solent. In this strait, two ocean tides that flow round Britain from the boundless northern seas meet in daily opposition off the mouth of the River Homelea.... and when the turbulence ceases, they flow back into the ocean whence they spring."

Bede's published writings found their way to corners of Europe as distant as Italy and Iceland, although sometimes in quirky and modified forms that harked back to earlier conceptions of what caused the tides. As expressed in the fourteenth-century Icelandic saga the Rimbegla, "Beda the priest says that the tides follow the moon, and that they ebb through her blowing on them, but wax in consequence of her movement." But the Rimbegla also offered quite a different mix of explanations. One was that "[at new moon] the moon stands in the way of the sun and prevents him from drying up the sea; she also drops down her own moisture. For both these reasons, at every new moon, the ocean swells and makes those tides which we call spring tides. But when the moon gets past the sun, he throws down some of his heat upon the sea, and diminishes thereby the fluidity of the water. In this way the tides of the sea are diminished." Another passage argued that "when the moon is opposite to the sun, the sun heats the ocean greatly, and as nothing impedes that warmth, the ocean boils and the sea flood is more impetuous than before — just as one may see water rise in a kettle when it boils violently. This we call spring tide."

Elsewhere, too, the growth in understanding of tides was anything but linear or consistent. By the time of the Arab geographer Edrisi (1110–1166), the Arab world extended to the Atlantic coasts of Spain and Morocco. Yet Edrisi described Atlantic tides without taking into account the daily retardation. This effectively severed the link between the passage of the Moon and the timing of tides. Thus, he attributed tides largely to the effect of winds. "The reason for [the rise and fall of the tides] is the wind which stirs up the sea at the beginning of the third hour of the day. As soon as the sun rises above the horizon the flood increases with the wind. Before the day ends the wind falls because the sun is setting and the ebb takes place," Edrisi wrote. "In the same way at nightfall the wind rises again and calm is not restored until the night draws to an end. The high tides occur during the 13th, 14th, 15th and 16th nights of the (lunar) month; then the waters rise unusually high, reaching a level which they do not attain again until the corresponding days of subsequent months"

Another Arab writer, El-Masudi, who died in 956, continued to hold views very similar to those of the ancient Indians. One was that the Moon warmed the bottom of the sea, which drew water out from the Earth; that water in turn expanded in volume and made sea level rise. Full moon brought the highest tides because it produced the most heat. Zakariyya ibn Muhammad ibn Mahmud al Qazvini (who died in 1283) added a more mystical explanation as well. "The agitation of the sea," he wrote, "resembles the agitation of the humours in men's bodies, for verily as thou seest in the case of a sanguine or bilious man, &c., the humours stirring in his body, and then subsiding little by little; so likewise the sea has matters which rise from time to time as they gain strength, whereby it is thrown into violent commotion which subsides little by little." He also proposed a religious cause of tides, writing, "and this the Prophet (on whom be the blessings of God and his peace) hath expressed in a poetical manner, when he say: 'Verily the Angel, who is set over the seas, places his foot in the sea and thence comes the flow; then he raises it and thence comes the ebb.'"

A wide range of contradictory ideas about tides continued to prevail in different parts of the world. Increasingly, it was in Norman Britain that the most accurate observations and predictions of tides were made. A table forecasting high tide at London Bridge, believed to have been written by Abbott John of Wallingford, was published in the early thirteenth century. Much like the Chinese tide prediction table for the bore on the Qiantang, it was simply a list giving the time of high tide for each of the thirty days in the lunar month (starting with new moon), and it took into account the daily retardation by making that time forty-eight minutes later each day.

Meanwhile, Gerald of Wales, a Welsh cleric who died around 1220, wrote a description of the tides in the Irish Sea that supported some of Bede's observations and advanced them as well. Comparing the times of tides along the British coast, he showed just how drastically they could differ over a relatively short distance. For example, when the tide was only halfway in at Dublin, he wrote, it was high water at Bristol. And when the Moon crossed the meridian, the tide was flooding at Dublin but at ebb on the northern coast of Scotland.

This meant that there was no shortcut to predicting tides. The relationship between the high tide and the Moon's passage of the meridian had to be determined by local observations for each individual port, such as for London Bridge on the Thames. (This would become known in tidal jargon as the establishment of that port.) How widely this relationship can vary is seen most clearly by comparing modern tide tables for places on opposite sides of the relatively narrow British Isles. A quick check on the Internet shows, for example, that the higher of today's two high tides at Liverpool on Britain's west coast comes about one half-hour past noon, whereas at Immingham on the east coast it will not be until about seven hours later. A similar situation holds true where I live in British Columbia. Vancouver Island is so long (in a roughly north-south direction), and it so severely obstructs the movement of tidal waters, that high tide occurs first on the outer (west) coast. But this high tide only reaches the east coast (the waters between Vancouver Island and the mainland) four to five hours later.

By late medieval times, at least some people with an interest in the tides recognized that they were linked in some way to the lunar cycle. But this acceptance was by no means universal. A whole range of other explanations continued to coexist and vie for attention with those that gave primacy to the influence of the Moon. There were mystical or vitalistic concepts, such as ideas about tides being among the Earth's own life functions. There were more naturalistic but non-astronomical theories, wherein tides were caused by the action of winds, or the heat of deep sea waters, or the effect of rivers flowing into the sea. And even when lunar cycles were presumed to be connected to the tides, the actual nature of that linkage was a mystery that would not be solved until the Renaissance.

Living by the Tides

2

I awoke that awful day by morning light to a loud staccato sound. The corrugated metal roof of my house slapped and rattled with each gust of wind. I was renting a simple, remote cottage, without electricity or indoor toilet, on Prevost Island, British Columbia. It was situated just above and behind the beach at the head of Diver Bay, a narrow body of water that was exposed to strong winds when they blew in from the southeast. And from the sound of it, this was a major storm. I went to the ocean-side window to check on my boat, which normally rode out storms quite comfortably at a permanent mooring in the middle of the bay. My heart sank at what I saw.

Seawolfe, my six-metre (twenty-foot) powerboat, should have been tossing in the waves directly in front of the house, in deep water well out from the beach. Instead, she was being driven into the shallows and was already more than halfway in to the shore. Her bow pointed into the wind and waves, as it should, and I could see the rope that still connected her to the mooring. But with each successive wave, the bow heaved and jerked on that rope. And with every heave, she drifted a little farther back toward the beach.

I could not see the gravel beach itself at all. The tide was so high that the waves were breaking right over the line of sun-bleached gray driftwood logs that usually rested high and dry, like a thigh-high berm of wood, along the back edge of the beach. The small plywood dinghy that I used whenever I needed to row out to *Seawolfe* lay there, upside down on the logs, just below the house. That was normally a secure place to leave it, but now the dinghy was getting battered by the waves. I slipped into my gumboots, raced outside, and went down to drag it to safety, right up onto the lawn. But there was nothing I could do about *Seawolfe*, except to stare in horror as her stern moved closer with every wave to that low barrier of logs.

I knew right away what had happened. I felt like a fool and cursed myself for my negligence. This was an unusually high tide, and it was being driven even higher by the storm surge that builds up when powerful winds push the sea in the same direction as the in-flooding tide. Storm surges can routinely add one metre (over three feet) to the normal height of a tide, and sometimes much more than that. It was ordinary high tides and winds that had carried those driftwood logs ashore and left them at the high, back edge of the beach in the first place. Now, with this extra-high tide, the sea was breaking much farther up on shore than at any time in the three years I had lived here.

Still, if the weight anchoring my boat's mooring to the bottom had been heavy enough, *Seawolfe* should have been perfectly safe. Unfortunately, the weight was inadequate. I had only recently bought *Seawolfe*, a hefty little boat with an enclosed cabin and inboard engine, and begun using her on my regular forty-minute crossing to the region's main village, Ganges, on Salt Spring Island. I went there once or twice a week for shopping, to get my mail, and to visit friends. Until then, I had been making that trip in a faster and much lighter open runabout powered by an outboard. The mooring I had put down in the bay was anchored to a couple of very large buckets of concrete, which was good enough for the light open boat. For *Seawolfe*, which was three or four times as heavy and had a high cabin structure that caught the wind, I knew I really needed to put down a heavier mooring. And I had intended to do so at the first opportunity. But time and tide are relentless and unforgiving. This high tide and storm surge arrived before I got around to what was just one more project on a long list.

And so I watched helplessly as *Seawolfe* waltzed her way sternward, lifting and dragging the mooring as she went. She was a sturdy, well-built wooden boat with a heavy, straight keel that could take a lot of abuse. But aft of the keel were two vulnerable features, the propeller and a cast bronze spade-shaped rudder that protruded downward and was not protected in any way. I could just imagine her drifting backward onto those logs and bashing that rudder against them as she tossed on the waves. The leverage exerted by the rudder might even tear a hole in her bottom. It was a terrible prospect.

Then, almost miraculously, an especially large wave swept in. *Seawolfe* rose high, jerked hard on the mooring, and was tossed up and onto the logs themselves. The next few waves bathed her bow in frothy foam and spray. But I could see that the keel now rested firmly across several large logs, while the rudder had hooked itself in behind one of them. She was stuck there. And it was a good thing, too.

Soon the tide began to fall. The wind remained strong, and the waves just as violent, but within an hour or so the seas were breaking lower on the beach itself and no longer sweeping over the logs behind it. It was safe for me to go down to where *Seawolfe* lay, nearly upright on the logs, and inspect her for damage. I was relieved to see that, aside from some scraped paint and shallow gouges in her bottom planking, she looked intact. The rudder shaft was slightly bent, but when I climbed on board and turned the wheel, it still swung smoothly to right and left. I checked the bilges for water and saw that she had not sprung any planks or otherwise begun leaking.

Still, I had only temporarily dodged the bullet. The tide prediction tables for the area gave the ominous numbers for that date. Although the tide was now dropping and would stay low most of the day, late afternoon would bring another high tide, in fact a slightly higher one. Meanwhile, the storm still raged. By evening, the sea would be just as high, and the waves would be churning around the boat again. She would pound up and down on the logs, this time perhaps for two or three hours while the tide was at its highest. Possibly worse, the waves and wind might well carry in some new, large driftwood logs that would bash against her sides. All I could do was to try and give her some protection against that horrid eventuality.

I hiked across the island to the home of my elderly landlady, Jean de Burgh, and her family, to tell them what was happening and to recruit help. The de Burghs owned almost the entire island and ran it as a sheep and cattle farm. One daughter, Barbara, and the hired farmhand loaded up a wagon with old tires and towed it with a tractor across the island and down to my place. We lashed the tires to the bow and sides of the boat with ropes, checked the time, and then waited nervously for the next high tide. It came in faster and higher than expected.

Meanwhile, we also noticed a favourable change. The worst of the storm had passed and the wind had eased. Much smaller waves were now breaking over the logs. Unlike in the morning, when I would not have dared to climb onto those logs in the midst of the seething tempest, in the late afternoon it was safe. Well before dusk, we realized that the boat might become buoyant while there was still some daylight left, which meant there was a chance to float her off. Barbara hurried home, switched on the generator, and used the radio-telephone to report the situation to our local volunteer search and rescue service — the private but non-profit B.C. Lifeboat Society. (The Canadian Coast Guard did not yet have a station in the Gulf Islands, as it does today.)

In twenty minutes salvation was on the way. An electrician named Horst Klein and his daughter Sandra, both wearing head-to-toe orange flotation suits, came roaring across from nearby Pender Island. They bounced over the waves in their Boston Whaler, a flat-bottomed and highly manoeuvrable fibreglass boat with two powerful outboards. Skimming in close to the beach, they hurled a light line with a loop and float on the end. I snared it with my long boat hook, pulled in the heavier towing rope that was tied to it, and attached the rope to the anchor bitts on *Seawolfe*'s foredeck. Then they idled just off the beach, waiting as the tide rose just a bit more, while I bided my time on board *Seawolfe*. Eventually, the boat began to rock and toss. She was afloat. The Kleins gunned their engines and headed away from the beach and toward the mouth of the bay. The towing rope went taut. *Seawolfe*'s rudder snagged once, then bounced and scraped its way over the logs. We were free. *Seawolfe* was still dragging the tires, which slapped against her sides and made her a dismal sight. But I did not care. My saviours throttled back their engines and towed us around to a sheltered bay on the other side of the island, where we anchored her for the night. The first chance I got, I wrote a cheque and sent a sizable donation to the Lifeboat Society.

The damage to *Seawolfe* was minimal, and I knew I had gotten off lightly. The worst bruise, in fact, was to my pride. I had let down my guard by showing insufficient respect for the tides and the power of the sea. And I should have known better. After all, I had been living by the tides for years.

I moved to British Columbia's Gulf Islands as an adventurous young guy in the early 1970s. Enchanted by the area, I made friends and settled first on scenic Salt Spring Island, a rugged evergreen haven of laid-back lifeways that lies just off the southeastern end of huge Vancouver Island. I still live there today. A large island with mountains, lakes, and fine harbours, Salt Spring is the most populous of a handful of relatively developed islands in the area. This means that they have paved roads, electricity, and telephone service and are accessible by car ferry. Salt Spring also has a high school, hospital, banks, stores, and restaurants; in short, all the amenities of civilization. There are also dozens of much smaller islands in the area, such as Prevost, with no such services. All the islands and their surrounding waters are blessed with a mild maritime climate and are sheltered by Vancouver Island from the winds and swells of the open Pacific. This makes the entire region — B.C.'s Strait of Georgia and the adjacent U.S. San Juan Islands and Puget Sound — a paradise for boating, fishing, and coastal living. By the time I arrived, the traditional economic mainstays of sheep farming, commercial fishing, and logging were making way for tourism, retirement homes, and summer cottages.

I brought with me a Klepper collapsible kayak, and the first thing I did was to poke around the local bays, beaches, and smaller islands. The area was ideal for someone like me, who enjoyed eating fresh shellfish. The rule was that you did not harvest clams or oysters in months without an "R" in them, i.e. during the summer. But when fall and winter arrived, the beaches were a cornucopia. In the Gulf Islands, winter low tides nearly always come at night. My friends and I would go out with buckets, shovels, and hurricane lanterns to dig clams in the soft mud of the estuaries and pick oysters off the rockier parts of the foreshore. As the northwest coast aboriginal saying puts it, "When the tide goes out, the table is set." There is even a B.C. folk song that repeats this lyric in its chorus. The abundance of sea foods extends to salmon, rock fish, sole, Dungeness crabs, edible seaweeds, even octopus, abalone, and sea urchins for those who take up scuba diving.

The Vancouver Island region, and Salt Spring Island in particular, had long attracted serious sailors, boat-builders, and wharf rats of all kinds. A surprising number of them had built their own boats and sailed around the Pacific, or even around the world. There was also a significant commercial fishing fleet. Inspired by what these people were doing, I, too, threw myself into the world of boats and boat building. Together with my friend Steve Phillips, who was a more skilled woodworker, I built an eight-metre (twenty-five-foot) wooden sailboat, *Lili Marlene*. She was such a masterpiece of cedar planking and varnished fir, mahogany, and spruce topsides and spars that people soon began approaching us to do alterations, repairs, and maintenance on their boats as well. For years, both of us earned much of our living from boat work.

As sailors, though, we were neophytes. We force-fed ourselves on all the obligatory books and magazines, but we picked up most of the essential knowledge, techniques, and experience the hard way, through hair-raising trial and error, sailing *Lili Marlene* into, and out of, one dodgy situation after another. We learned how to handle the sails, read and interpret charts, set an anchor, and steer a compass course even in fog or heavy rain. (The boat had no radar, and this was before the days of GPS navigation.)

Learning to cope with the region's large tides and powerful tidal currents was a major part of the challenge. Where I lived, on the southern B.C. coast, the maximum tidal range was about four metres (thirteen feet), which put it just into the lower end of the macrotidal range. Farther north along the coast, however, the tidal range got much larger. One time I went by ferry to Prince Rupert, which is just south of Alaska and where the tidal range can reach eight metres (twenty-six feet). I was a foot passenger and walked right off the ferry. But the tide was so low that the overhanging rear end of a tour bus hung up on the steep ramp as it tried to drive off. This blocked all the other cars, which had to wait nearly two hours for the tide to rise before they could disembark.

Soon after launching *Lili Marlene*, I moved from the comforts of Salt Spring and took up living on smaller and more remote islands. This required getting back and forth in my own boat, and the tides became a much greater factor in my everyday routine. First, I grabbed at the chance to be the winter-season caretaker of tiny Russell Island, just off the southern end of Salt Spring. The owner, a California dentist named John Rohrer, loved beachcombing. All sorts of interesting stuff drifted in on the tide there, much of it highly useful. For many years, Rohrer had been salvaging driftwood logs, squared-off heavy timbers, and creosoted wood pilings, using them to build things like heavy retaining walls for paths and gardens. And he had devised a clever system that took advantage of the tides to ease the burden of this task.

Somehow Rohrer accumulated dozens of old, worn-out bicycle tire inner tubes, which he could tie and string together like a chain of huge rubber bands. When he found a valuable log or timber lying on the shore, it was often well out on the intertidal zone and too heavy to move up to the high tide line by hand. But if he left it there, the next high tide and breeze might carry it away again. The solution was to tie one end of a chain of inner tubes to the log, stretch it tight, attach a rope to the other end of the tubes, and tie the rope to a tree or boulder above the high tide line. Whenever a higher tide came in, the elasticity would ease the log farther up onto the intertidal zone. And the beauty was that it would do so automatically, without anyone having to be there, even if that tide came at night or in miserable weather. When Rohrer went away for the winter and I took over as caretaker, he assigned me the task of patrolling the shores for useful flotsam and applying his system. And it worked like a charm. By the time he came back the following summer, I had salvaged a dozen or more choice logs and timbers. All were lying securely at the highest tide line and lashed by ropes to trees, so they would not float away and could be further dealt with at leisure.

Next I moved to a more long-term living situation on Prevost Island, a much larger privately owned agricultural island that was farther from Salt Spring. There was a sheltered bay on one side of the island, with a dock where the de Burghs could keep their boats and bring in a barge to deliver heavy supplies or to take their farm animals to market. But the cottage that I rented from them at Diver Bay was too exposed to the prevailing southeasterlies to build a pier and floating dock. Waves and huge, heaving driftwood logs would have destroyed any such fixed structure. I had no choice but to keep my boat out in the middle of the bay at a permanent mooring.

Everything I did depended on the tides. During the years when the lightweight open runabout was my commuting boat, if the weather was calm I could briefly nose it into the shallows in front of the house to load or unload, wearing high gumboots to keep my feet dry. But because the tide was usually either rising or falling, I could not leave it there for long. The only safe and convenient place for it was out at the mooring. When the much heavier *Seawolfe*, with its deeper keel, became my commuting boat, it was hardly ever practical to bring her into the shallows. I reached her by rowing out in a small dinghy that was light enough to store above the high tide line. When I went off in my powerboat, I usually left the dinghy behind, tied to the mooring.

The whole business of coming and going could be a bit tricky and tedious, especially at night by flashlight, or when the tide was very low and I had to drag the dinghy out through an oozing mass of mud and eel grass before I reached deep enough water to climb in and row. But some of my work was right there on Prevost Island, helping out on the farm or doing maintenance on the de Burghs' boats and buildings, so I only went to Salt Spring every few days. In stormy weather, especially in winter, I was sometimes pinned down for a week at a time, unable to go to "town." It was something I simply had to accept.

But it was a magical place to live. The bay, lined with tall firs and gnarly red arbutus trees, was totally secluded and private, without any neighbours within sight or earshot. I could put down a trap and catch a Dungeness crab right in front of my window, or jig a sole for dinner and grill it outdoors on a wood fire. I revelled in the wildlife on land and sea. Herons nested by the dozens in one wooded area on the island. Eagles perched in lofty snags overhanging the shore. Playful harbour seals and river otters would nose around the dinghy and follow me, diving and cavorting, when I rowed around the bay. Minks prowled the driftwood that lined the beach.

And the incessant rise and fall of the tides added immeasurably to the charm and excitement. The great variation over the course of an average day provided a nice rhythm and made the bay an ever-changing and always intriguing place. At extreme low tide, the exposed intertidal zone was so extensive that it gave the bay an entirely different look and feel from the situation at high tide. At low tide, the shore was never silent. There was always the snapping sound of exposed seaweed popping in the sun, the spurt of filter-feeding clams deep in the muck shooting up water through their siphons, the mewing of impatient, hovering gulls, or the cracking of a shell as a predatory gull or raven swooped and nabbed a red rock crab. There was also a pleasantly pungent smell, a rich mix of iodine from the kelp and other simmering seaweeds and sulfur from the decay of worms and other critters that had lived in the mud. When the tide was in, the air was cleaner, fresher. At those times, I enjoyed the lapping of the sea as it licked at the logs high up on the strand, or the swash of waves coming in and out as they ground away at the gravel of the beach and at bits of wood and other flotsam.

At certain times of the year, if the tide and winds were just right, thick rafts of dead seaweed would drift in and be left at the tide line to dry in the sun. Stuck together in long matted rows, it was very easy stuff for me to gather. I simply rolled it up, carried it to my vegetable garden, and applied it as fertilizer. The tides also delivered to my doorstep count-less small pieces of driftwood, which also dried in the sun. Much of it was so broken up by wave action that I could collect and use it without further cutting or splitting. It was full of absorbed salt, and I dared not use it indoors in my wood heating or cooking stoves. The salt would have destroyed the metal stoves and chimney pipes. But it was perfect for burning in the outdoor fire pit, where in nice weather I often grilled a freshly caught fish.

Just as at Russell Island, high tide at Diver Bay left a steady sup-ply of large logs and milled boards or timbers on the beach. Most of it consisted of commercially harvested logs that had broken free from log booms that were being towed to sorting grounds or directly to mills. Under the law, only licensed commercial log salvagers were allowed to patrol the region's beaches in their boats and remove these logs, which they sold to a consortium of the big logging companies. In reality, though, the salvagers were only looking for the very best of the lost logs, such as large-diameter cedars and Douglas firs that had not been in salt water too long. These licensed beachcombers frequently came and went, leaving behind lots of smaller logs, including many that were a godsend for a variety of my projects, and nobody would know, or care, what hap-pened to those logs. When I wanted to enclose my garden, for example, it was a snap to gather fence posts from the great selection of driftwood on the beach.

Bob Storey, a friend of mine living on Salt Spring Island, decided to add a room onto his small seaside cabin. I agreed to help him salvage the wood he needed from nearby beaches. He was fussy and wanted to build the addition exclusively out of perfectly straight Douglas fir logs of a consistent diameter. We went out three times in my runabout, for a couple of hours on each occasion. Selecting carefully, Bob was able to find a few dozen excellent logs, which we towed directly to his home. The only cost to him was some gasoline and the requisite bottles of beer to lubricate the pleasant proceedings.

The tides were also very helpful when it came to boat maintenance. Every boat that stayed for months on end in sea water accumulated a growth of barnacles and slime below the water line. This had to be scraped off at least once a year. Then, the clean boat bottom had to be washed with fresh water and given a coat of anti-fouling paint to help retard the future accumulation. Most years there were also minor repairs that needed looking after. This could all be done at a commercial boat-yard by pulling the boat out of the water on a marine railway, but that was expensive. For me, it also would have meant a two-hour trip each way to get to the nearest boatyard. As a low-budget wharf rat, my better option was to take care of these boat-bottom chores by careening my boat, as mariners have been doing for millennia. Each year I ran the boat into the shallows an hour or so after high tide and grounded it there, right in front of my house. When the tide dropped, it rested, secure and dry and leaning over on one side. This usually gave me at least five or six good hours in which to scrape and clean half of the bottom. Then I slopped anti-fouling paint on that side of the bottom before the tide came back in again and eventually floated the boat off. The following day I could do the opposite side.

I also put the large tides at Diver Bay to good use after that near disaster, when *Seawolfe* dragged her mooring in the storm surge. I had to assemble a much heavier mooring and sink it in just the right spot in the bay. A permanent mooring consists, first of all, of a very heavy weight that sits firmly on the bottom. Coming up to the surface from that weight is a length of chain (or a combination of chain and rope) that attaches to a large float with a convenient ring on top. Often there is also a floating rope line with a loop in it. This makes it easy to pick it up (even in the dark) with a boat hook and put it over the boat's mooring bitts. At that time, many people in the islands were using old car or truck engine blocks as inexpensive bottom weights. But to bring one of these out to my remote site and drop it accurately would require a large boat or barge, and perhaps a powerful crane of some kind, plus the boat's owner or crew to run it. This could get expensive.

Fortunately, there was a do-it-yourself method that made use of the tides. First, I waited for a very low tide. This was important, because I had to be sure to place the mooring far enough out from the beach that the boat would always have enough water under the keel to toss up and down without hitting bottom, even at the lowest tide of the year. But I did not want it to be farther out from the beach than necessary, because that simply meant a longer row each time in the dinghy. Rowing around the bay, I plumbed the depth with a lead weight on the end of a rope until I found a spot that was as close as possible to shore but still had sufficient depth. There I dropped a cinder block with a long piece of lightweight rope tied around it holding a small, temporary float to mark the target.

Next, also at low tide, I rolled a big empty oil drum, with one end re-moved, out onto the tidal flat. Using a few sacks of Portland cement and gravel from the beach, I mixed wheelbarrows of concrete and gradually filled the drum, which stood upright in the mud. While the concrete was still soft, I anchored one end of a length of very heavy salvaged chain into the concrete, using old bolts and pieces of rebar set crosswise to make sure the chain could never be pulled out of the concrete. To the free end of the chain, I shackled one end of a longer piece of very heavy nylon rope and attached the other end of that rope to my surface float with shackles and swivels.

By the time I was finished, the tide was already creeping back in, and I had to abandon the heavy mooring contraption and allow the con-crete to set. But there was no need to worry. It was not going anywhere. The next part of the job involved moving it to the spot I had chosen and marked, which was farther out in the bay and where the water was deeper. During the next daytime low tide, I assembled a crude raft and tied it to the top of the drum of concrete. The raft consisted of a ladder-like framework of scrap wood with large pieces of salvaged styrofoam lashed to it. (Like logs, styrofoam, which was widely used as flotation for docks, often drifted in on the beach.) Then I waited for the tide to come in, float the raft, and lift the drum off the bottom. At that point, it was easy to tow the whole shebang out to my predetermined mooring site. All I had to do after that was lean out from my dinghy, cut the rope tying the raft to the drum, and jump back out of the way. The raft flew up in an explosion of water as the drum of concrete plummeted to the bottom, with the chain, rope, and surface float all attached. Voila! A safe mooring heavy enough to hold *Seawolfe*, *Lili Marlene*, or any other boat I was likely to own.

During my years at Diver Bay, I began sailing *Lili Marlene* north for a few weeks each summer to explore. On the B.C. coast, heading in that direction meant getting into increasingly remote areas and also into bays and channels with larger tides. In some of those waters, the coastal mountains were so rugged and steep that there were hardly any good places to make a clearing and build a house. Quite a few people lived in floathouses, homes built on a raft of logs or a barge that could be tied up next to a steep shoreline, where they rose and fell with the tide. Or floathouses could be beached on gently sloping mud flats. In other spots, ordinary wood-frame houses had been dragged off barges (usually on a framework of heavy log skids) and placed in clearings just above the beach. All of these approaches had the advantage that the dwellings could be moved easily and inexpensively to a new place, depending on changing circumstances. This was important to some of the gypsy-like small-time loggers and commercial fishing families, for example when the children of those families reached school age and had to be within striking distance (by boat) of a larger settlement.

Everything about daily life in those areas had to be synchronized with the tides. This has been captured beautifully in a memoir by an old-time B.C. logger, trapper, and fisherman named Bill Proctor. *Full Moon, Flood Tide* depicts coastal life, mainly from the 1930s to the 1970s, as experienced by the people who lived and worked on the bays and inlets of B.C.'s central and north coast. Even today, as Proctor says in his introduction, "The most significant factor affecting the lives of people who live by the sea is the association of the big flood tides with each full moon. Innumerable activities of both human and animal residents are timed to these high tides," beginning with the movements of the prime fish species, the Pacific salmon. "When salmon are migrating home to their natal rivers and streams they always come in from the ocean on the full moon tides, therefore, this is the time for the best fishing."

Proctor shows how the various types of fishing boats exploit the Moon and tides in different ways. For example: "Gillnetters [which catch salmon by snagging them in long, lightweight nets with openings that just fit the size of their gills] love the full moon tides for two reasons. The first reason is the obvious fact that more fish are around. The second reason is that when the moon is full, the phosphorescence in the water is dimmed so the fish cannot see the net as they can on the dark or new moon."

The Native peoples of the area have always been people of the sea. "For thousands of years," Proctor continued, "the aboriginal people of the coast waited for the big tides, for this was when they caught their fish. Herring always spawn on the peak of the full moon tides and eulachons [an oily fish that was a staple for many coastal tribes] always enter the rivers on the full moon tide." Likewise, low tide was when "you could dig clams or gather abalone, mussels, crabs and barnacles. When the tide is low and stands still just for a short time, it is known as low water slack. The low water slack at the time of the big tides is the best time to fish halibut or cod and sometimes big spring salmon."

Nor was food gathering the only activity that depended on tides. Towing logs was always easier when done with a flooding or ebbing tide, rather than against it. So was towing a floathouse. "The hand loggers [who slid their cut logs downhill from the steep coastal slopes to the beach] made use of the flood tide for getting logs off the foreshore. A-frame loggers [who used engines to pull logs out of the woods] used the high tides for getting equipment onto the foreshore. It was always a lot easier to heave cables and blocks in under the trees on a high tide than to pack them up a slippery beach [at low tide]." All the practical logistics of coastal life were carefully timed, usually to coincide with a high tide. "A high tide is the time to put a boat on the ways and often to let it off. If you want to move a house from land onto a float, or from a float onto the land, the full moon tide is the time to do that. If a propane tank, or a generator or other big piece of machinery needs to be delivered, you know it will be coming in on the high tide."

I, too, always needed to know the exact timing and predicted height of the tides, which were printed in an annual booklet published by the Canadian Hydrographic Service. Never was this more obvious than one particular summer in the early 1980s. Usually, when I went sailing, I was accompanied by friends. But on this occasion, I was alone and found my-self heading south in the middle of the Strait of Georgia, the main body of water between Vancouver Island and the B.C. mainland. I had spent a long day cruising along in very light breezes, and not getting anywhere fast. But late in the day the wind suddenly came up strongly and from the southeast, which meant right in my teeth. Beating to windward, I struggled to make headway and gradually realized that I was going to be caught out in the open strait by approaching darkness. There was simply no safe and sheltered anchorage that I could reach in the hour or two of light that remained.

With a companion on board, we could have continued on sailing through the night, alternating watches to go below, get warm, and catch bits of sleep. But I had never spent a night out on an exposed body of water sailing alone, nor was my boat properly equipped to do so. Strong winds and rapidly building seas made the prospects much worse. I was not at all confident that I could handle such tasks as reducing sail alone and in the dark, with just a flashlight in one hand to provide illumina-tion. I was afraid that, in trying to do so, I might get hit and injured by the mainsail boom, or even knocked overboard. And I was sailing smack in the middle of the main shipping lane between Puget Sound and Al-aska, which was plied by cruise ships, freighters, and other large vessels. They had radar, and I had a radar reflector hanging in the rigging. But mine was a wooden boat that might not show up well on their screens, and my running lights were only kerosene lanterns. So there was the real danger of being run down in the dark.

The closest shore was a large island called Hornby. There was no acceptable harbour on the eastern side of the island, which faced me. But a closer look at my chart revealed a little oblong nook on Horn-by, a cove with a narrow entrance and a single number printed in the middle, which represented a sounding. That number was a one, which meant that, at least somewhere in that nook, there should be water with a depth of roughly one fathom (exactly six feet, or nearly two metres) at low tide. This was not really quite enough, since my deep keel drew almost that much, and wave action tosses a boat up and down. The cove was so small that it did not even have an official name, and the chart gave no other details, just the rough shape of the shoreline and this one indi-cator of depth. Normally, I would never have considered sailing into an anchorage that was so shallow and for which I had so little information. But this was not a normal situation.

So, I got out my binoculars. From what I could see, the entrance was just wide enough to sail through, and I could not make out waves breaking on any obstructions. Nor was there a kelp bed or other indicator that an unmarked rock might be lurking in that narrow entrance. Then I looked at my booklet of tide tables, and eureka, I saw a glimmer of hope. The tide was presently quite high, which greatly improved my chances of getting into the cove safely. Best of all, the table showed that, although the tide would drop during the night, it would not drop all that far. It was the kind of low tide that should, I reckoned, still leave a little water under my keel. And then it would rise again to another really high tide in the morning. This was the diurnal inequality (or pattern of mixed tides) that is so typical of the northwest coast. There was a high tide, followed by a period of relatively high low water, then another high tide, and only after that a genuine, get-out-the-gumboots low tide. But that really low water would not come until early the following afternoon. In the meantime, I should be able to get into that tight anchorage, spend the night, and slip away again in daylight the next morning.

I decided to give it a shot. I put the helm over, sheeted in the sails, and picked up speed, reaching across the wind. If the conditions had been calm, I would have motored slowly up to the mouth of the cove and taken a good look before entering. But I only had a small outboard that hung off the stern. In any real waves it was worse than useless. So I had to rely on the wind, and I was sailing in faster than I would have preferred, but at least I had a fallback position. As I got close enough to see into the cove, if there was any kind of visible hazard, the wind direction was favourable for a quick turnaround. As long as I brought my boat about in time, I could safely turn and sail away from the shore.

I closed quickly with the island. What had looked like toys from a distance rapidly grew into houses. I could make out a few small boats pulled up on shore inside the cove, and someone rowing around in a dinghy. The waves in the open strait swept along parallel to the Hornby Island shore and curled around into the entrance of the cove without generating any telltale froth or white water that would indicate shallow rocks. There were very small patches of bull kelp floating on the surface along both sides of the entrance, but not in the middle. And the centre of the cove, where I was pointing my bow and intended to anchor, looked perfectly clear. I was committed.

I sailed in at a good clip. The waves abated as soon as I passed through the slot. In seconds I reached the middle of the cove, where I spun the boat ninety degrees, until my bow was pointing into the wind and my sails were slatting. Leaving the tiller, I scooted forward, heaved my anchor off the bow, and waited anxiously to see if the hook would bite. It did. Quickly lowering my sails, I breathed a sigh of relief and looked around. It was a lovely spot, a tiny, idyllic body of water with a half-dozen cute summer cottages lining its well-treed shores. The wind was still howling out in the strait, and waves burst over a low line of rocks. But in behind those boulders, there was only a little choppiness that rocked *Lili Marlene* gently.

I sat down in the cockpit to relax and thought about going below to get myself a beer, but before I did I saw a man rowing swiftly out from shore. I hailed him as he approached. Although his expression was friendly, he looked perturbed as well. "This is not a place where you can anchor your kind of boat," he told me after a few pleasantries. "It dries right out at low tide. No one brings deep-keeled boats into here. See," he added, pointing to the southern end of the cove, "only small boats are kept in here." I explained to him that, according to the tide table, there was not going to be a really low tide during the night, and that I planned to leave at high water in the morning. He still looked skeptical and bobbed around alongside, waiting to see what I would do.

As a backup for the outboard, I had a very long oar, or "sweep," lashed to the deck. I sometimes used it to row *Lili Marlene* slowly under dead-calm conditions, usually just to get into or out of a dock. I untied the oar and stuck it down into the water to measure the depth exactly. It showed that I had a little over one metre (about four feet) of water under my keel, and the tide would be dropping about that far during the night. So, the local fellow was right. I might well hit bottom as the tide ebbed. "Is there any place in here that's a bit deeper?" I asked him. "Well, yes. Over there, where those skiffs are moored. When the tide's right out, there's still a little pocket of water over there." Okay, then. I told him I'd move and anchor there. He still shook his head, but I explained that with darkness coming on fast, I could not face heading back out into the strait alone, especially with it blowing so hard. He wished me luck and rowed home.

I fired up the outboard, pulled up my hook, motored over to the little cluster of skiffs and dinghies, and anchored as close to them as I could get. Then I plumbed the depth with my lead line and saw that the guy had been right. It was a good deal deeper than the original spot and would have to do. At last I had a chance to relax and enjoy a beer. I was damp with sweat from the nerve-wracking experience. I stripped down and dove in for a quick and refreshing dip. Later, as I started to cook dinner, I heard a voice hailing me from the closest spot on shore. It was a different guy. He asked me where I was from and whether I would like to join him and his wife for dinner. I thanked him but said that I already had my meal almost prepared. "Well, why not come in afterwards for some pie and coffee, then?" And so I dinghied ashore after dark and spent the evening unwinding and enjoying their hospitality.

During the night, tucked into my bunk, I could hear the waves crashing on the nearby rocks and the wind whistling through my rigging. But I was snug and safe in the marginal little anchorage. The tide dropped quite a bit. By moonlight, when I got up to pee over the side, I could see that much of the cove, including the entrance, was dry, but there was still water around the cluster of skiffs. I poked down with my oar and found that there was just barely enough water under my keel. Still, it *was* enough. By morning, the wind was down and the tide was back up. I dared not dawdle. The waves out in the strait were small enough that I decided to rely on the outboard. As soon as it was light, I weighed anchor and motored slowly out of the cove.

Not until I was well away from Hornby Island could I truly relax, raise my sails, and head off in a light breeze. I offered silent thanks for my modern charts, tide tables, and the fortunate combination of tides that had saved my bacon. Two more good days of sailing and I would be home.

Heaven and Earth

The Renaissance and Isaac Newton's Great
Breakthrough in Tidal Theory

"See those bronze lions' heads?" said our guide over the tour boat's sound system. I looked over to the stone wall of the Embankment, just below Waterloo Bridge on the Thames. Two green tarnished lions' faces were set into the wall about one metre (just over three feet) below the level of the famous riverside street. At the moment, the murky surface of the Thames was lapping along the wall about one and a half metres (four or five feet) below those heads. "When the water is up to their mouths, we say that the lions are drinking. That means it's a really high tide."

Along with a few dozen fellow visitors, I had paid my return fare and boarded the small catamaran for a one-hour boat ride down the Thames from near the Parliament buildings at Westminster to Greenwich in southeast London. The September weather was sunny and warm, a perfect day for a river cruise. No matter that much of the river traffic consisted of tugboats moving garbage scows full of dumpsters. The guide announced one historic attraction after another: the Grapes Pub, where

Charles Dickens sang for customers as a child; trademark Tower Bridge; a modern reconstruction of Shakespeare's Globe Theatre; the Canary Wharf development with its Canada Tower, the tallest building in Britain; a replica of Sir Francis Drake's ship, *Golden Hinde*. And there were the endless docks and warehouses, where goods used to be unloaded by hand from ships arriving from all corners of the world during Britain's heyday as a maritime and trading power. Today, the container terminals are elsewhere, so most of these old brick waterfront buildings have been converted into outrageously expensive riverside condos with balconies, flowers, and fabulous river views.

The other passengers may have been lazily sightseeing, but for me the cruise was like a pilgrimage. This particular stretch of the lower Thames, with its large five-metre (sixteen-foot) tides, was long at the forefront of both tidal research and maritime history. It is no coincidence that the earliest tide tables in Britain were for London Bridge, where knowing the tides was so crucial for captains during the days of sail. Regardless of the wind, heading downriver on an ebbing tide was usually the quickest and safest way to get beyond the Thames estuary and out to sea. London was where key intellectual breakthroughs were made in tidal science, especially by a group of learned men associated with the Royal Society. Greenwich, with its docks and observatory in close proximity, was where some of the earliest systematic measurements of tides were taken in the seventeenth century. And Isaac Newton, the greatest single theorist of tides, once took a boat ride down the river to consult the Astronomer Royal at Greenwich.

As a lover of wooden boats, I was delighted when the first thing I saw after getting off at Greenwich was the beautiful nineteenth-century tea clipper *Cutty Sark*, one of the fastest tall ships that ever sailed. Now a maritime museum, she was resting in a tidal drydock and looked to be in perfect condition, ready to sail again at any time. On a high tide, the drydock had only to be flooded and the gates opened and she could be towed onto the Thames and out to sea. I went on board and paced the long wooden deck, staring up at the rigging and imagining myself at the huge wheel as she heeled and pranced in a westerly blow. Sadly, since my visit, she has been severely damaged by fire.

I strolled through the grounds of the National Maritime Museum, with its huge anchors and buoys, and up a manicured grassy hill to the

Royal Observatory, with its grand, if antiquated, telescopes. This was where observations had been made to create star charts for the Royal Navy's navigators. A fine collection of old chronometers was housed in glass display cases. These little devices had made it possible, for the first time, for explorers like Captain Cook to determine longitude easily and accurately. Dozens of schoolchildren were posing and taking one anothers' pictures while straddling the prime meridian. This is the north-south line through Greenwich that marks the world's zero point of longitude (and is the baseline for Greenwich Mean Time, as indicated by a line across the observatory's courtyard and a very large digital clock). Hanging in the halls were portraits in oil of the long series of distinguished Astronomers Royal. I noted especially Sir George Airy, who made major contributions to the study of tides during the nineteenth century.

It was fitting that, by the time I got back on the boat a few hours later for the return ride upriver to Westminster, the tide had risen. When we passed the Embankment again near the end of the trip, the river was lapping just below the green leonine faces on the wall. And the tide was still rising. The lions were about to take a drink.

Copernicus, Kepler, and Galileo

The Renaissance of the sixteenth and seventeenth centuries was a time both of unique geographic discovery and great change in Western thinking about the natural world, including the tides. Yet it was far from being a rapid and unidirectional revolution; change was gradual, halting, and incomplete. The period spanned the transition from medieval to modern times, when rival ways of thinking coexisted uneasily or clashed overtly. Church dogma and obeisance to principles and "received" learning that had been passed down from the ancients, especially Aristotle, coexisted with a new focus on measurement, experimentation, and the rapid sharing of knowledge in an ongoing process of learning.

These contradictions were manifest in the lives of that era's greatest scientific heroes. Galileo Galilei discovered the moons of Jupiter, but he was later forced by the Inquisition to recant his views. Newton worked out the mathematics of universal gravitation, but he was also a secretive alchemist who nearly poisoned himself on mercury fumes while seeking the transmutation of base metals to silver and gold. Johannes Kepler formulated the laws of planetary motion, but he was driven from one country to another by religious conflict, was himself excommunicated, and had to defend his mother against accusations of witchcraft.

Probably the greatest single break with the past, and of particular relevance to understanding the tides, began when the Polish Nicolaus Copernicus (1473–1543) challenged the long-accepted Ptolemaic view that the Earth was the centre of the universe, surrounded by a nested series of perfect crystalline spheres bearing the Sun, Moon, planets, and stars. Copernicus proposed instead that the Earth and other planets all revolved around the Sun, but he did not work out the details.

Johannes Kepler, a German who had trained as a mathematician and then devoted his efforts to astronomy in Prague, was the father of celestial mechanics. He discovered and formulated the laws of motion that governed the Copernican system. A few years before the telescope was invented, he measured and calculated the orbit of Mars and determined that it was not a circle but an ellipse with the Sun at one of its two foci. The orbits of the other planets and the Moon were ellipses as well. And whereas in the Copernican system the Sun was at the centre but played no apparent role in the movement of the planets in their orbits around it, Kepler proposed that these planetary motions must be the result of some real physical force that emanated from the Sun.

He was uncertain what that force might be, but his description of it hinted at something very much like gravitational attraction, and he saw it as governing the movement of the Moon around the Earth as well. In his *Astronomia Nova*, published in 1609, Kepler argued that there must be a force keeping the Moon in its orbit around the Earth, and that this force might also account for the tides. Based on a calculation that the mass of the Moon was 1/54th that of the Earth, he wrote: "If the Moon and Earth were not retained by animal force, or some other equivalent force, each in its orbit, the Earth would ascend toward the Moon a fifty-fourth of their distance, and the Moon would fall toward the Earth by 53 parts of this distance, whereby they would be united ..."

Applying this force to the oceans, he added: "If the Earth ceased to attract [to itself] the waters of the sea, they would rise and pour themselves over the body of the Moon." And the way he saw this producing tides was as follows: "The sphere of the Moon's drawing power extends to the Earth and incites the waters under the Torrid Zone [the tropics] … insensibly in enclosed seas, but sensibly where there are very broad beds of the ocean and abundant liberty for reciprocation of the waters."

The following year, Galileo gazed for the first time at Jupiter through his newly built telescope and saw it being orbited by four large moons. These provided a convincing model in miniature of the Copernican scheme for the solar system as a whole. Kepler read about Galileo's discovery and responded with enthusiasm. He, too, soon looked at Jupiter's moons through a telescope and wrote "Conversation with the Siderial Messenger," a long and laudatory letter of support for Galileo's celestial observations and ideas. In another tract, he defended Galileo against his critics.

But the Italian would not accept the German's ideas about what was not yet being called gravitation, and mocked him in particular for his notion that the Moon might be causing the tides. Galileo could comprehend and accept direct mechanical causes, such as the winds blowing over the sea and building up waves, but he simply did not believe in the possibility of action-at-a-distance, a process in which one body could influence another across an empty void.

To Galileo, it smacked too much of medievalism and magic. "That concept is completely repugnant to my mind," he wrote scathingly. "I cannot bring myself to give credence to such causes as lights, warm temperatures, predominance of occult qualities, and similar idle imaginings." Singling out Kepler for his wrath, he added: "I am more astonished at Kepler than at any other…. Though he has at his fingertips the motions attributed to the Earth, he has nevertheless lent his ears and his assent to the Moon's dominance over the waters, to occult properties and to such puerilities."

Galileo would have none of it. He proposed, instead, a complex and primarily mechanical explanation of tides. He had noticed that if he held a vase of water and moved it in a jerky, irregular way, the water in the vase also took on this motion. He had seen the same thing happening on a larger scale in the barges that brought fresh drinking water to the city of Venice. The accelerations and decelerations of the container, with its sudden movements, caused the water to slosh back and forth in a wave-like pattern that, he thought, resembled the rise and fall of the tides. So he tried to explain by reference to purely mechanical causes how such regular and repeated oscillations might be set in motion in the Earth's oceans.

In a private letter to Cardinal Orsini in Rome in 1616, Galileo drew a diagram of the Earth circling the Sun over the course of a year and also rotating each day on its own axis. He analyzed these motions, trying to imagine how they might affect the behaviour of water held within a particular sea or ocean basin on the Earth's surface. For part of each day, he argued, that body of water would be rotating in the same direction as the Earth was revolving around the Sun in its annual orbit. But twelve hours later, that sea or ocean basin would have moved to the opposite side of the Earth, and it would be moving in an opposite direction to the Earth's orbit. In one case the two motions would be added together, resulting in an acceleration, while in the other case one motion would be subtracted from the other, resulting in a deceleration. He also thought that particular seas or ocean basins probably responded to such disturbances by oscillating according to their own natural periods (just as the barges full of drinking water seemed to do), which thereby produced regular and repetitious tides and also accounted for the observed differences in tidal patterns from one ocean or sea to another. Galileo's concept fit reasonably well to the pattern of two tides a day. However, it did not do a very good job of accounting for the way the timing of tides seemed to follow lunar movements, or how they varied in size over the course of the month and year.

Britain's Royal Society and Isaac Newton

In Britain during the first half of the seventeenth century, the concept of gravity gained gradual ascendancy as a way of explaining how and why the heavenly bodies followed curved paths. As Kepler had reasoned, there seemed to be some force of attraction holding the Moon and planets in their orbits.

John Wilkins, a clergyman and teacher who was to be instrumental in founding the Royal Society (which would prove to be active in tidal studies throughout the coming centuries), was one of the first to think and write about the nature of attraction toward the centre of celestial bodies. In 1638, he published *The Discovery of a World in the Moone*, a tract that argued for the existence of life on what he thought was an eminently habitable satellite (complete with an "Atmo-Sphere") that circled our world. To Wilkins, the Moon seemed to be held in its orbit by a force very much like gravity, although he did not call it that.

Since the Moon "moves so swiftly as Astronomers observe, why then does there nothing fall from her, or why doth shee not shake something out by the celerity of her revolution?" he asked. "I answer, you must know that the inclination of every heavie body, to its proper Center, doth sufficiently tie it unto its place." And the consequences of raising an object above the surface of the Moon seemed clear. "Suppose any thing were separated [from that heavy centre], yet must it necessarily returne again, and there is no more danger of their falling into our world then there is feare of our falling into the Moone...."

Wilkins thought all heavenly bodies held sway over such objects out to some distance from their centres, and within that circumference anything raised into the sky (above the Earth, for example) "would fall downe to us." Furthermore, "Though there were some heavie body a great height in that ayer, yet would the motion of its centre by an attractive virtue still hold it within its convenient distance."

He also imagined how earthly gravitation — although he did not explicitly call it that — might influence a bullet that was fired high into the sky above the rapidly rotating Earth. He concluded that it would follow a trajectory "by reason of that Magneticke virtue of the center (if I may so speake) whereby all things within its spheare are attracted with it."

The exact nature of such an attractive force was still very much an open question. Even earlier, in 1600, the prominent English physician William Gilbert (Queen Elizabeth I's personal doctor) had demonstrated that the entire Earth was actually a giant magnet. His book, *De Magnete*, was the first to describe how the Earth, spinning on its axis, generated a powerful magnetic field — this is why compasses work as they do — and to show the relationship between magnetism and electricity. He did not believe that the attraction of magnetism was magic, but the language of the times still smacked of the occult. It was an "orb of virtue," for example, that surrounded a magnet and extended out from it in all directions. Not surprisingly, he suggested that it was a magnetic attraction that linked the Earth and the Moon. Since by this time it was widely acknowledged that the timing of the tides seemed to follow the Moon, he proposed that this attraction also caused the tides.

Even two generations later, the nature of these attractive "virtues" remained maddeningly unclear. Then, in 1663, John Wallis, an Oxford professor of geometry who was prominent in the Royal Society, made a very prescient observation that Newton would also adopt. Whereas Kepler had focused on how smaller bodies orbited much larger ones (such as the Earth circling the Sun and the Moon circling the Earth), Wallis pointed out that it is actually the combined centre of gravity of the Earth and the Moon that orbits the Sun. By "gravity," though, he did not imply universal gravitation, as understood by Newton, but simply the centre of weight or mass of the two bodies, which were somehow (and still more than a bit mysteriously) linked to each other.

"How the Earth and Moon are connected, I will not undertake to shew," he wrote, "but that there is somewhat that doth connect them (as much as what connects the Loadstone [i.e. a magnet] and Iron) is past doubt, to those who allow them to be carried about by the Sun as one aggregate body whose parts keep a respective position to one another."

As for the tides, Wallis thought the regular rise and fall of the sea "has so great a connection with the Moon's motion, that in a manner all philosophers have attributed much of its cause to the Moon." And this operated "either by some occult quality, or particular influence which it has on moist bodies, or by some magnetic virtue, drawing the water towards it, which should therefore make the water highest where the Moon is vertical." To him, the case was so clearly proven that "it would seem very unreasonable to separate the consideration of the Moon's motion from that of the sea."

At the same time, across the Channel on the megatidal shores of France, the philosopher René Descartes offered a theory or "system of the world" that was entirely different from any that relied on magnetic or other attractive virtues to hold the heavenly bodies in their orbits. Adopting an ancient concept, he proposed that out beyond the atmosphere, space, far from being empty, was filled with invisible matter, a medium called aether, or ether. All the motions of celestial bodies could thereby be explained (in a mechanical way that probably would have appealed to Galileo) as the result of rotational currents, or vortices, set up in this ether by the large body at the centre of each region of space. Thus, for example, as the Sun turned on its axis it created a rotational effect in the ether surrounding it that, in turn, swept the planets around with it. Likewise, the Earth's rotation generated a vortex that carried the Moon around in its orbit. To account for the tides, Descartes reasoned that as the Moon orbited the Earth, it compressed the ether between its orbit and the Earth's surface. Where the Moon passed over the sea, this compression of the ether also pressed down on the atmosphere, and in turn on the sea's surface, causing the tides.

Descartes's scheme was acclaimed as an intellectual tour de force, especially in France. Like Galileo, he was reluctant to accept vague and unexplained notions of action-at-a-distance. But in England the increasingly popular empirical approach was able to discredit this scheme. Christopher Wren (the architect who designed St. Paul's Cathedral) proposed an experiment to test Descartes's theory. The idea was simply to use a barometer filled with water to see whether air pressure varied in accordance with the movements of the Sun or the Moon across the sky. There proved to be no such correlation, and Isaac Newton was among those who were quick to point this out.

Newton, born in 1642 in Lincolnshire, was a strange and reclusive genius who never married and had almost no friends during the first half of his life and his career as Lucasian Professor of Mathematics at Cambridge. (This same chair has been held in recent years by Stephen Hawking.) He mainly kept to his rooms at Trinity College for days on end and neglected his meals and personal grooming. He was so secretive that he invented calculus but did not reveal it for decades, until forced to when Gottfried Wilhelm von Leibniz in Germany published a similar mathematical system. And he pursued personal disputes with a spirit of vituperation and petty spite.

Once Newton's genius was recognized, though, he moved to London, became active in the Royal Society (eventually serving as its president), sat in Parliament, and befriended such luminaries as the diarist Samuel Pepys and the philosopher John Locke. He was knighted and became Master of the Royal Mint, a post concerned with revamping Britain's coinage and cracking down on counterfeiting, which he did with a vengeance, insisting that the culprits be executed rather than merely rot away in the Tower. He died a wealthy man and was interred at Westminster Abbey. Voltaire, who was in London at the time, marvelled that a natural philosopher and commoner should be given a funeral fit for royalty.

Newton's intellectual system may have encompassed the solar system and universe, but his actual life transpired along a very narrow stretch of English soil between inland Lincolnshire, Cambridge, and London. The entire route was within a day's journey of the sea, even by horse-drawn carriage. Yet ironically, although we know that he took at least one boat ride along the Thames, the person who contributed most to the world's understanding of tides probably never set eyes on the open sea itself. Moreover, the theory of tides, expressed in dry mathematics, occupied only a few pages in his greatest single work, the *Principia Mathematica*. Written in Latin and published in 1687, this dense tome was not translated into English for decades. It was only by way of a short synopsis, drafted initially by his friend Edmund Halley to give King James a thumbnail sketch of the theory, that the world gradually came to understand Newton's explanation of gravity and tides.

Newton created a greatly simplified model of how the Earth, Moon, and Sun move in relationship to one another and act on one another gravitationally. He used it to tease out the most essential elements of what causes tides on Earth.

Stripped of the mathematics, his approach postulates and envisages an ideal and hypothetical planet Earth that is a perfect sphere with no continents, only a very deep and uniform ocean, an envelope of water covering its entire surface. Of great significance for the tides, that sphere also has a relatively massive moon orbiting around it. (With the sole exception of Pluto and its moon Charon, our moon is much larger, in proportion to the mass of its parent planet, than any other satellite in the solar system. The Moon's mass is about 1.2 percent that of the Earth.)

Newton realized that it is wrong to see the Moon as simply revolving around the Earth while the Earth remains fixed at a central point, or even — to be more precise — at one focus of a slightly eccentric ellipse. One of his greatest insights was that gravitation is a *mutual* force, and a universal one. In other words, each massive body in the cosmos attracts every other body. And following on the earlier suggestion made by John Wallis, this meant that, even though the Earth is so much larger and more massive than the Moon, both of them actually orbit around a point that represents their common centre of mass, also called the barycentre.

This imaginary and transitory spot — and the motion around it — may best be visualized by thinking of the two bodies as being linked to each other by a long and rigid rod that extends out through space, a bit like a cheerleader's baton, with a large, heavy ball on one end and a much smaller and lighter ball on the other. If such a baton were twirled in the air, the lighter end would trace out a large circle, but the heavy end would trace out a small circle as well, revolving in a sort of wobble or pivoting motion.

In a similar way, the Earth and the Moon constitute a single stable system in which they are linked to each other by their mutual gravitational attraction. The system is stable in the sense that they neither fly apart nor fall in upon each other. And so, over the course of each lunar month, both bodies revolve in a slow and graceful celestial dance around that common centre of mass. Because the Earth is roughly eighty times as massive as the Moon — Kepler's estimate of fifty-four times was wrong — this centre is an imaginary point located about 1,600 kilometres (1,000 miles) deep within the Earth and on whichever side happens to be facing the Moon at any given time.

The Earth's radius is about 6,400 kilometres (4,000 miles), which means that the common centre of mass of the revolving Earth-Moon system is a point located about 4,800 kilometres (3,000 miles) out from the Earth's actual physical centre, and it lies along a line between the centre of the Earth and the centre of the Moon. (See Figure 3.1, which also shows the vertical axis around which the system pivots.) It is important to bear in mind, however, that since the Earth itself rotates on its own axis every day, the centre of mass is not an actual, fixed physical point in the interior of the Earth. It is only an imaginary one that always lies at the same distance along that line, even as the physical Earth turns. But because this point, and the axis around which the Earth-Moon system rotates, is quite far from the Earth's actual centre, most of the Earth's mass is always concentrated on the side farthest away from the Moon as the entire Earth-Moon system pivots as a unit around this common centre of mass.

It is, in other words, an eccentric, or off-centre, system. The Earth's mass is heavily weighted out toward one side as the system turns, just as the heavier ball of the cheerleader's baton traces out a small circle in the air when the baton is twirled. This has consequences, because when

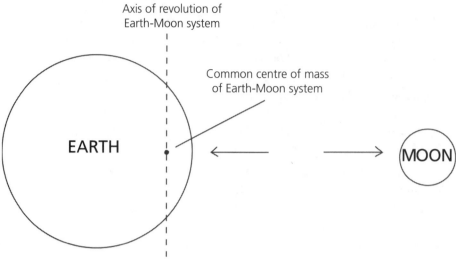

Figure 3.1: The revolving Earth-Moon system.

anything is twirled in the air, such as a weight on the end of a string, the person twirling that weight experiences centrifugal force. It feels as though the weight is trying to fly away from the twirler. Likewise, as the Earth revolves around the common centre of mass during the lunar month, there is a centrifugal force acting on the Earth and all of its mass equally. (Every part of the Earth is revolving in a monthly circuit around the barycentre with a radius of about 4,800 kilometres, or 3,000 miles.) This force acts on every portion of the Earth, whether made up of solid matter, such as rock, or of liquids, such as ocean water. And this centrifugal force always points directly away from the Moon.

Now, the revolving Earth-Moon system is a stable one that is held together in a fixed configuration even as it spins. This means that some equal and opposite force must be maintaining a balance, just as the string keeps the weight from flying off and away. That force is, of course, the gravitational attraction between the Earth and the Moon. This exactly counterbalances the centrifugal force.

But Newton had an additional brilliant insight. He saw that, although the entire Earth-Moon system seen *as a whole* remains in balance (with gravity exactly equalling centrifugal force) there can be an imbalance in these forces as they affect any *particular* part of the Earth's mass when it is looked at on its own. After all, gravitation acts between every unit of mass in the cosmos, and it varies in strength according to their distance apart. (In fact, it varies in inverse proportion to the square of the distance between them.) It does not act *only* between the Earth and the Moon taken as entire entities. It makes just as much sense to look at the gravitational attraction between, say, a bucket of water at a particular place on the Earth's surface and a clump of rock in the centre of the Moon.

So, let's look at the gravitational attraction exerted by the Moon (in this case by that rocky satellite considered as a whole) on a particular portion of the Earth's mass (say a unit of water on the surface of the ocean). The strength of that attraction will be slightly different depending on whether the particular unit of watery mass is on the side of the Earth facing the Moon (i.e. closer to the Moon) or on the side facing away (i.e. farther from the Moon). And since the imaginary deep ocean on Newton's hypothetical Earth is fluid and able to move, it is this *difference* in gravitational pull that leads to the tide-generating (or tide-raising) forces.

On the side of the Earth facing away from the Moon, the gravitational pull of the Moon on the ocean will be a bit weaker than its average gravitational pull on the Earth as a whole. But centrifugal force exactly balances this average gravitational pull. On the side of the Earth facing away from the Moon, therefore, centrifugal force will be slightly stronger than the gravitational force, and this extra net force will be exerted on any particular unit of ocean water in a direction pointing away from the Moon. That unit of water will tend to move away from the Moon and "pile up" a bit, creating a bulge.

Similarly, on the side of the Earth facing toward the Moon, the gravitational pull of the Moon on the ocean will be a bit stronger than its average gravitational pull on the Earth as a whole. On that side, gravity will slightly exceed centrifugal force (which is acting in a direction away from the Moon) and will exert a net force that pulls the ocean waters toward the Moon. The water will bulge out on that side as well. The result, Newton reasoned, is a tendency for the envelope of deep water surrounding this ideal Earth to take on a symmetrical double bulge with its alignment toward the Moon and away from it. (Scientists today describe this shape as a prolate spheroid.)

The simplest such situation to consider is one in which the Moon's orbit (which is tilted relative to the Earth's equatorial plane) has taken the Moon to a point on a line that lies along the plane traced out by

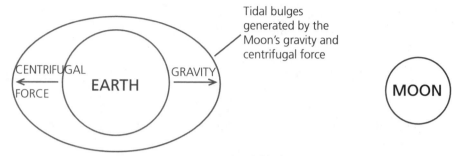

Figure 3.2: Newton's tidal bulge.

the Earth's equator. (See Figure 3.2, which greatly exaggerates the bulge of the liquid envelope.) But now imagine looking down on this same system from a spot far out in space along the Earth's axis of rotation and therefore high above the North Pole. (See Figure 3.3.) And imagine that there is a very high volcanic mountain, much like one of the islands of Hawaii, that pokes up from the deep ocean floor and through Newton's imaginary deep ocean on or near the equator. As the Earth rotates rapidly on its own axis during the course of a day (a motion quite distinct from the much slower revolution of the entire Earth-Moon system as it swings around its common centre of mass each month), that island will first pass through one of the bulges and experience a raised water level, or high tide.

Then, roughly six hours later (see the pointed form representing the mountain), as the rotation continues, it will pass through an area of the ocean midway between the two bulges, where the sea level is considerably lower. This would be experienced on the island's shore as a low tide. Six hours later again, the Earth's rotation would carry the island through the second bulge, for another high tide, and after a further six hours or so, another low tide. In the course of a typical day, therefore, this isolated volcanic island would experience two episodes of relatively high water and two of relatively low water. There would be two similar complete tidal cycles each day, or what scientists call a semi-diurnal tide.

The actual situation is slightly different, though, because of the monthly revolution of the Earth-Moon system, which occurs in the same direction as the Earth's rotation on its own axis. By the time the Earth has rotated once, the Moon has also moved a significant way along in its orbit, carrying the double bulge in the ocean with it, so that a typical full

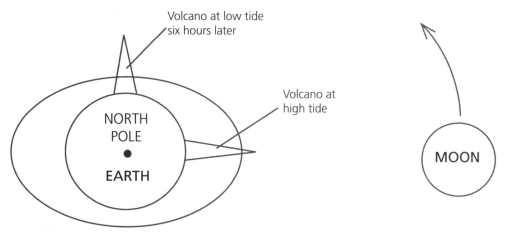

Figure 3.3: Changing water level on shores of volcano.

tidal cycle takes not twenty-four hours but approximately twenty-four hours and fifty minutes. But the basic principle is valid. And Newton's theory accounted well for the observed phenomenon (especially around most of the shores of Britain, although not everywhere on Earth) that the prevalent tidal pattern was two nearly equal high tides and two nearly equal low tides each day.

There was another very important element to Newton's theory. He realized that the Moon was not the only heavenly body generating earthly tides. The gravitational pull of the Sun does the same thing. The Sun is much farther away from Earth, however. So, despite its vastly greater mass, its effective gravitational pull on the Earth is less, roughly 46 percent as strong as the Moon's. Still, this is significant, and it can be imagined as creating its own set of tidal bulges in the Earth's ocean. However, since the Earth's ideal Newtonian ocean is a single body of water, the hypothetical bulges caused by the Moon and Sun must either augment or counteract each other, depending on the positions of the Moon and Sun in the Earth's sky at any given time.

For example, consider the situation where, for a few days each month, the Moon is on the same side of the Earth as the Sun. (On Earth, we see this as a time of new moon.) Astronomers call this conjunction. Half a month later, the Moon will be on the opposite side of the Earth from the Sun, and we on Earth will see this as a time of full moon. Astronomers call this opposition. And both situations are sometimes referred to as being in syzygy.

During these periods when the Moon and Sun are in syzygy, the lunar and solar bulges they generate will add up to particularly large cumulative bulges. This produces large spring tides (which have nothing to do with spring as a season), which occur roughly twice a month. (See Figure 3.4.) On the other hand, when there is a half moon in the sky, it means that the Sun and Moon are *not* in alignment. In fact, their gravity is tugging at the Earth at right angles to each other. In this situation, called quadrature, the Sun's tide-generating force is trying to create bulges on areas of the ocean that are midway between the larger bulges generated by the Moon, and so their effect is essentially subtracted from the tidal effect of the Moon. The result is smaller-than-average total bulges, and smaller overall tides, called neap tides. These also occur roughly twice a month. Here again, even though he based his logic on a drastically simplified picture of the cosmos, Newton's theory accounted quite well for some of the ways tides appear to work in the real world.

By Newton's time, however, European mariners had begun to discover places in the world where the tides did *not* follow the pattern that prevailed around the shores of northern Europe (and on the northeastern coast of North America as well), where there were usually two roughly equal high tides a day, and two roughly equal lows. On certain

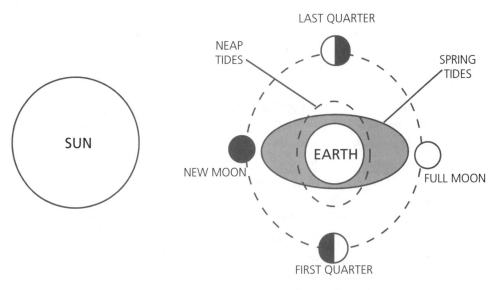

Figure 3.4: Spring and neap tides.

shores (such as on the northwest coast of North America, from California to British Columbia to Alaska) one of the highs (or lows) each day was usually found to be much higher (or lower) than the other. Today, this pattern of mixed tides is called a diurnal inequality.

But Newton was able to explain how the diurnal inequality could occur as well, at least in principle and to a good first approximation.

The double tidal bulge shown in Figure 3.2 fits the situation where the Moon is at a point on the Earth's equatorial plane. But more often, because the Moon orbits with a significant angle of declination, the Moon is at a point well to the north or south of this equatorial plane, as in Figure 3.5. The double bulge caused by the Moon's gravity will in that case also be aligned along that angle of declination, with one bulge predominantly to the north of the equator and the other to the south. Now, imagine that the Moon is high in the sky of the northern hemisphere, and picture our volcanic island poking up through the ocean at the latitude of the Tropic of Cancer, roughly the situation of the Big Island of Hawaii.

In the course of an earthly day (which also includes the night, of course), as the Earth rotates that island would experience first a high tide as it passes through the large bulge created on the side of Earth facing

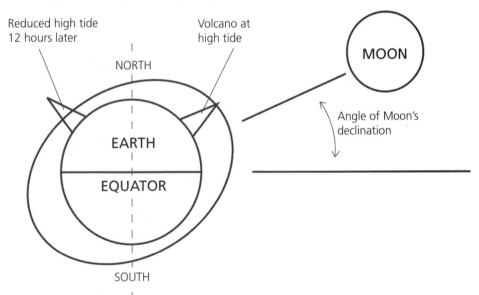

Figure 3.5: Tidal bulge with the Moon orbiting at an angle of declination.

the Moon. Six hours later, the island will pass through the "non-bulge" at the usual place between the two lunar bulges and experience a low tide. But then, six hours after that, it will *not* pass through a second large bulge, as it did when the Moon was on the plane of the Earth's equator. Rather, it will pass through a region of the ocean where the lunar bulge is only a bit greater than in the "non-bulge" regions. In short, there will be only one really high tide in roughly twenty-four hours, with a much smaller second high tide a half-day later. This second high tide may be only very slightly higher than an ordinary low tide and barely distinguishable from it.

There are many other complexities in the way that the tide-generating forces behave. For example, the declination of the Moon does not usually match the angle of the Sun relative to the Earth's equator. Even when the Moon and the Sun are in syzygy, i.e. in approximate alignment relative to the Earth, they do not usually exert their tugging forces along the exact same line. There are times when the Moon is much higher in the sky than the Sun, and vice versa. Their relative positions vary over the course of each month and year and from one year to the next. Moreover, the distances between the Moon and Earth and between the Earth and Sun also vary in regular but long cycles. All of which means that the tide-generating forces vary significantly over time as well.

Just as important is the fact that the Earth does not have an ocean of uniform depth, or one that is free to move with the tide-generating forces, unobstructed by continents. So the way the tide-generating forces affect the Earth's actual oceans is far more complicated than in Newton's model. As European mariners continued to explore the world's shores, and as people managing ports and directing navies took systematic and increasingly accurate measurements of tides, it gradually became apparent that not even Newton's theory could explain the entire range of tidal phenomena. But he had identified some of the key elements in the basic celestial geometry that leads to earthly tides. It was an unprecedented breakthrough in human understanding. For a few decades there were holdouts, especially in France among the supporters of Descartes. Ultimately, though, all serious efforts to refine our knowledge of tides began with Newton's insights into the tide-generating forces and progressed from there.

Into the Maelstrom

Powerful Tidal Currents Generate Killer Whirlpools

It was a horrible sight to behold, and the dull roar of the tumult was amplified by the steep wooded hillsides that lined the channel. A stone's throw out from shore, frothy bands of white water raced around in a large circle, swirling, pulsating, gyrating. In the middle was a sloping hole in the sea as wide as a house. Belching and gurgling, the ocean was being sucked down and swallowed whole.

My view of the churning whirlpool was just a bit too close for comfort. I was holding on tight to the seat and cockpit coaming of a small boat. It swayed and oscillated as it was swung by the currents like a toy along the outer edge of the violently wheeling ring of water. Adrift without power, the boat began to tilt in ominously toward the centre of the vortex. Staring down into the foam spinning around the whirlpool's cone-shaped maw, I thought to myself, *What a stupid way to die.* And I had only myself to blame.

The whole thing started as something of a lark. It was the mid-1970s, shortly after I first settled in British Columbia and before I built

Lili Marlene. I was still pitifully ignorant of seamanship and coastal conditions. My friend Daryl had just bought a funky old wooden boat about six metres (twenty feet) long with a tiny cabin. The boat seemed seaworthy enough, without obvious leaks or rot. But it was also heavy and sluggish, especially when powered by a decrepit Johnson outboard that could only push us along at around four knots.

Daryl and his girlfriend Kirsten had decided to motor north up the Inside Passage, the mainly sheltered network of channels that extends from Puget Sound in Washington State through British Columbia and on up into the Alaska Panhandle. They invited me to come along. Their plan was to settle at Hornby Island — the same place where, years later, I was able to duck into that tiny anchorage for a night on board *Lili Marlene* — and they had brought along their meagre personal belongings. Strapped to the top of the cabin were bicycles and camping gear. This was the tail end of the hippie era, and with our long hair and slightly tattered clothes, we looked like a waterborne gypsy caravan, and were treated that way by the redneck owner of a marina where we had to gas up and spend a night along the way. As for me, I was just along for the ride.

The first inkling of trouble came as we approached one of the tidal choke points that make navigating the Inside Passage tricky. It is called Dodd Narrows, where the currents race through at up to nine knots. We had a copy of the tide and current tables and knew we had to be careful at such spots. We planned to reach the narrows around slack tide, the only safe time for a slow boat like ours to transit the passage. After that, the tide would begin to flood northward. But we did not realize just how short the period of slack could be, and arrived an hour late. While we were still a half-hour south of the narrows, a tug pulling a raft of logs passed us in the opposite direction. Tugs towing log booms always wait for dead slack to negotiate such channels, so we realized we had missed the ideal time. But at least we would not be bucking the current, which would be impossible with our underpowered boat. The tide, we knew, would be with us, carrying us along. So what was there to worry about? We never even considered turning back while we still had the chance and waiting for the next period of slack. On we went.

Seen from a plane, Dodd Narrows is a roughly hourglass-shaped gap between huge and mountainous Vancouver Island to the west and small, low-lying Mudge Island to the east. As we came up into the tapering southern end of the hourglass, we passed high cliffs on the Vancouver Island shore topped by straight, tall Douglas firs and twisted arbutus trees with their peeling bark. The Mudge Island side had much lower forested terrain. At the tightest spot, the distance between the shores was only around 100 metres (328 feet), and the cliffs on the western side made it feel even narrower.

The tide was already flooding northward with strength when we reached that point. Daryl throttled back, but we still found ourselves being sluiced through the gap at a remarkable speed. The rocks and trees on both sides of the channel sides shot past. Where the tidal race was pinched against the shore, it kicked up whitecaps and doubled back into powerful eddies. The sea became ever more turbulent. Little staccato waves danced and jolted us. Our boat heaved in agony, its planks groaning in protest. The outboard motor exaggerated every movement. Hanging off the stern, it plunged deep one moment then pulled up the next. Each time this happened, the propeller bit in futility at the air. It revved and screamed. The channel ahead was clear, though, with no boats coming our way and no obstructions, so I still figured we were safe.

Then, when we were a little beyond the narrowest point and into the widening opposite end of the hourglass, there it was, lurking straight ahead. The whirlpool. The first I had ever seen. From a distance, it did not look all that threatening, just a low, swirling feature in the chaotic cataract of salt water. And the channel was now easily wide enough for evasive action. Daryl steered us off to one side of the rotating ring of spume, trusting the engine and current to carry us rapidly past.

But the outboard, tortured by the heaving and bucking of the boat, had other ideas. Suddenly it gasped and fell silent. Above the squawking of the gulls, which were feeding all around us on fish and crustaceans brought to the surface by the tidal race, we could hear the thrumming convulsions of the water. In seconds, our bow began to swing over. Drifting and fishtailing and sliding ominously sideways without the power needed to steer, we were drawn helplessly toward the outer rim of the burbling, cascading dervish.

Now we could see how determined the spinning leviathan was to have its way with us. We found ourselves slipping inexorably into the kind of nautical nightmare that has given mariners sleepless nights for centuries and stimulated some great flights of literary fancy.

"'Maelstrom! Maelstrom!' they were shouting. The Maelstrom! Could a more frightening name have rung in our ears under more frightening circumstances? Were we lying in the dangerous waterways off the Norwegian coast? Was the *Nautilus* being dragged into this whirlpool?"

So begins the climactic episode of Jules Verne's nineteenth-century science fiction classic *20,000 Leagues Under the Sea*. Captain Nemo's fantastic submarine is about to confront the world's most infamous tidal cataclysm. This rapidly rotating ocean gyre occurs at a spot just off the remote Arctic coast of Norway, a place where the regular rise and fall of the tide is translated by an unusual configuration of the seabed and surrounding shore and islands into swift and incredibly powerful horizontal currents. Over the centuries, these have sunk many ships and drowned hundreds of people.

There are other kinds of ocean currents caused by differences in water temperature, salinity, or prevailing winds. We know them by names like the warm Gulf Stream off the eastern United States, and the cold Humboldt Current, which sweeps up the western coast of South America. But none of them run at anything near the speed of the fastest tidal currents. And most flow quite steadily in a single direction, whereas tidal currents generally reverse direction twice a day.

"As you know," Verne's tale continues, "at the turn of the tide, the waters confined between the Faroe and Lofoten Islands rush out with irresistible violence. They form a vortex from which no ship has ever been able to escape. Monstrous waves race together from every point of the horizon. They form a whirlpool aptly called 'the ocean's navel,' whose attracting power extends a distance of fifteen kilometers [over nine miles]. It can suck down not only ships but whales, and even polar bears from the northernmost regions."

Verne goes on to paint a picture of a sea gone mad, with the *Nautilus* swept around crazily in a "spiral whose radius kept growing smaller and smaller," its steel plates beginning to crack under the strain. His novel ends with a tantalizing question mark hanging over the submarine's ultimate fate.

Verne was not the first writer whose imagination was captured by the awesome power and violence that is generated in the Norwegian

Arctic by swift tidal currents. Two decades earlier, Edgar Allan Poe had penned his gripping short story "Descent into the Maelstrom." Like Verne, Poe indulges in a bit of literary licence, especially regarding the sheer size of the whirlpool. But he also includes a precise description of how it is created by tidal action, as seen by his narrator, who sits on a craggy cliff overlooking the Lofoten Islands and their turbulent waters: "I became aware of a loud and gradually increasing sound, like the moaning of a vast herd of buffaloes.... At the same moment the chopping character of the ocean beneath us was rapidly changing into a current which set to the eastward. Even while I gazed, this current acquired a monstrous velocity. Each moment added to its speed — to its headlong impetuosity. In five minutes the whole sea ... was lashed into ungovernable fury ... heaving, boiling, hissing — gyrating in gigantic and innumerable vortices."

These wheeling waters soon combined and "took unto themselves the gyratory motion of the subsided vortices, and seemed to form the germ of another more vast. Suddenly — very suddenly — this assumed a distinct and definite existence, in a circle of more than half a mile in diameter." Inside this ring of foam there formed a deep depression in which the water rotated "dizzily round and round with a swaying and sweltering motion, and sending forth to the winds an appalling voice, half shriek, half roar, such as not even the mighty cataract of Niagara ever lifts up in its agony to Heaven."

Poe then has an old man tell his narrator an appalling story. One time, he and his brother were out on the water in a small open fishing smack and were swept by a storm right into the jaws of the Maelstrom itself. "Never shall I forget the sensations of awe, horror, and admiration with which I gazed about me," the old man said. "The boat appeared to be hanging, as if by magic, midway down, upon the interior surface of a funnel vast in circumference, prodigious in depth, and whose perfectly smooth sides might have been mistaken for ebony."

The tongue-biting agony continues. "Our first slide into the abyss itself, from the belt of foam above, had carried us to a great distance down the slope." But gradually they slipped even farther with each great turn around the whirling funnel, deeper and deeper into the liquid chasm. "Our progress downward, at each revolution, was slow, but very perceptible."

At last, Poe engineers a miraculous escape. The old man thought about the kinds of flotsam he has noticed washed up on nearby shores, and realized that only relatively spherical objects have the buoyancy needed to survive intact the whirlpool's fatal attraction. "I no longer hesitated what to do. I resolved to lash myself securely to the water cask upon which I now held, to cut it loose ... and to throw myself with it into the water." He tried to signal to his brother to do the same, but the latter shook his head despairingly and refused to budge. "It was impossible to force him; the emergency admitted no delay; and so ... I resigned him to his fate, fastened myself to the cask ... and precipitated myself with it into the sea."

Soon he witnessed his brother's dreadful demise. "Having descended to a vast distance beneath me, [the smack] made three or four wild gyrations in rapid succession, and, bearing my loved brother with it, plunged headlong, at once and forever, into the chaos of foam below." But the buoyant barrel saved the old man as the tidal cycle ran its course. Gradually, the whirlpool lost its power, and he found himself floating on the calm ocean surface at slack tide. "Speechless from the memory of horror," he was rescued by friends in a boat. "But they knew me no more than they would have known a traveler from the spirit-land. My hair, which had been raven-black the day before, was as white as you see it now."

A wonderful story, but how true is it to reality? Are there really whirlpools so large and powerful that sizeable boats get sucked down into them and disappear?

Because of its remote Arctic location (more than 67° north latitude), the Norwegian Maelstrom has been less thoroughly studied by oceanographers than other stretches of water in the world that are torn by strong tidal currents. Wildly differing claims have been made about the speed of the currents that produce the gigantic whirlpool. The most credible sources indicate a velocity of around nine knots, which is certainly fast, but not as fast as many other tidal currents. Nor is the area's 3.5- to 4.5-metre (12- to 15-foot) tidal range, which is similar to that prevailing along the southern coast of British Columbia, especially large. As for the size of the whirlpool, it is probably something like 40 to 50 metres (130 to 165 feet) across. Although nothing like the half-mile diameter claimed by Poe, this still makes it an incredibly large and dangerous vortex that mariners would approach only at great peril. But as we shall see, there are others of comparable size and power.

What makes the Maelstrom special is that it occurs in an area of relatively open sea, as opposed to within a constricted basin or narrow channel. The key is the topography of the local seabed. The Maelstrom forms near the southern end of a chain of islands, the Lofotens, that angle toward the mainland of Norway, forming a funnel shape that is open to the sea on its southwestern end. Between the Lofotens and the mainland is a body of water, Vestfjorden, that is roughly 500 metres (1,640 feet) deep in places. According to one modern estimate, as the tide rises and falls, 370 million cubic metres (97 billion U.S. gallons) of water have to sluice their way into and out of that deep, funnel-like fjord every six hours. And connecting the farthest islands along the chain is a relatively shallow submerged ridge with an area 45 to 75 metres (150 to 250 feet) deep, through which much of that water is channeled by the inexorable propagation of the tide. It is this vast volume of fast-flowing water — it runs deep and is then suddenly forced upward and accelerated over this shallow ledge or sill — that creates the spectacular gyre.

Thus, a simple vertical sea level change generates a horizontal flow, which then engenders a complex concoction of mighty vertical, horizontal, and rotational effects. A sea gone mad indeed. But different bodies of water go crazy in different ways.

Swirling Waters in Japan and Scotland

Most of the world's powerful tidal currents occur not on the open sea but in narrow channels, or where the tidal range is extremely large, or where both factors combine to augment each other. In Japan, the Naruto Whirlpools wreak their havoc in the Naruto Strait, a channel slightly over a kilometre (almost a mile) wide between the islands of Awaji and Shikoku, where vast amounts of water flow in and out between the open Pacific and the very large Inland Sea. Large freighters are routinely swept out of control there by the whirlpools and vortices and end up running aground on the rocky shore.

Another very dangerous series of whirlpools, the Corryvreckan (from a Gaelic word meaning "cauldron of the speckled sea," but locally

Illustration by Hiroshige of the Naruto Whirlpool in Japan.

called "the old hag") occurs in the Hebrides of western Scotland between the islands of Jura and Scarba. There the situation is a narrow strait one kilometre (over half a mile) wide combined with tidal currents that run at around nine knots. The British Admiralty's official *Pilot* describes "heavy overfalls" that "extend as much as three miles [five kilometres] seaward," with "very violent and dangerous turbulence." The guide to mariners advises that "no vessel should attempt this passage without local knowledge." As in the case of the Maelstrom, the currents gain strength because of a sudden rise in the seafloor that catapults masses of deep flowing water to the surface. In the middle of the channel, a large, conical pinnacle of rock reaches to within about 60 metres (200 feet) of the surface, compared to a depth of more than 180 metres (600 feet) for much of the surrounding water.

This pinnacle is a favourite spot for scuba diving during the short periods of slack, but many divers have perished because they did not leave the area in time as the tide began to run with strength. British scientists conducted an experiment there by placing a human dummy into the vortex along with a device that recorded depth. It was sucked down straight to the bottom and came up some distance away bearing traces of gravel from the sea floor. Any living diver, they concluded, would almost instantly be killed by the rapidly changing pressure during the rush to oblivion in the depths.

Many more lives have been lost there over the centuries in boating disasters. Author George Orwell nearly became a victim in 1947, when he took his young nieces and nephews out fishing in an open boat without paying due attention to the tide and current table. As nephew Henry Dakin told a local newspaper, Orwell "seemed to know what he was doing.... But as we came around the point obviously the whirlpool had not receded. The Corryvreckan is not just the famous one big whirlpool, but a lot of smaller whirlpools around the edges. Before we had a chance to turn, we went straight into the minor whirlpools and lost control." Orwell was at the tiller, but he was unable to steer. "The boat went all over the place ... pitching and tossing so much that the outboard motor jerked right off from its fixing." Orwell begged his much younger and stronger nephew to grab the oars and row. This saved the day. "We got close to a little rock island and as the boat rose we saw that it was rising and falling about twelve feet [almost four metres]." Dakin jumped off into the shallows, holding the boat's painter and hoping to pull the boat to safety. Just then, though, the heaving sea turned the boat right over, throwing the other occupants out. The imperturbable Orwell surfaced holding another nephew, only three years old, in his arms, and they all scrambled ashore.

British author Simon Winchester once hired an old local lobster fisherman to take him out for a look at Corryvreckan. "It was a terrible, magnificent thing," Winchester wrote. "Many had died in it, the old man said. His own brother was one such, who had drowned in the whirlpool thirty years back, coming home from a fishing trip." As they approached, Winchester suddenly saw "a ragged line of white. Breakers. Spray. And, faint at first but growing above the throb of the diesel engine, a steady growling roar." The boatman pulled his sou'wester down over his face and gripped the wheel tightly. "We could be in for a little wetting," he told Winchester with a look of mischief. "It's the perfect tide for it. A flood tide and a westerly gale. Ideal, if you know what you're about."

The boat began to rear and pitch, "seemingly quite beyond the control of the wheel. Currents began to tear her off her course, and the rocks … began to race by, now on the starboard side, now ahead, now to port, now astern, as we wheeled around, caught by forces unseen, deep, hugely strong."

The boatman pointed to an oily-looking area of sea about thirty metres (one hundred feet) across. Winchester "gazed at the flatness until, without warning, a plume of contorted and confused water belched upward, then fell back upon itself and began in a matter of seconds to swirl into a vortex — a real, perfectly formed whirlpool, thirty feet [nearly ten metres] across, with us perched delicately on its edge."

The boatman opened up the throttle and edged away from the whirlpool, only to encounter another oily circle of water that soon erupted as well. Steering away, they came upon another, and then another. For at least an hour, "we reared and plunged in this manner, and the waters boiled and bubbled around us." Then, "just as suddenly as it had begun, it was over," and everything was calm. Winchester got off lightly compared to many people who have experienced such an intimate rendezvous with tidal fury.

Spanish Explorers Encounter British Columbia's Whirlpools

The fastest tidal currents in the world occur along the coast of British Columbia. Because the area is readily accessible (at least compared to the Norwegian Arctic), they have also been the most thoroughly studied by oceanographers using sophisticated modern current meters. Large whirlpools are generated at many spots where fast tidal streams flow through the narrow entrances to long, deep fjords (locally called inlets) or through tight channels between the region's many islands.

The first experience of the whirlpools by European mariners came in 1792, when two Spanish captains, Dionisio Alcala Galiano and Ceyetano Valdes, were charting a portion of the Inside Passage. Their small Spanish schooners, the *Sutil* and *Mexicana*, came upon the mouth of a treacherous narrows that is now called Arran Rapids. The two ships' commanders went out in a longboat to have a preliminary look at the situation and reported seeing "such violent whirlpools that the water sank more than a yard [one metre]."

But it was the sheer speed of the current that most impressed them. "The extraordinary swiftness the waters acquired was a phenomenon worthy of the greatest attention." They estimated its speed at around twelve knots, or faster than what they had encountered in the Strait of Magellan, which was already notorious for its terrible tidal currents. "The aspect is a most strange and picturesque one," they wrote. "The waters look like a race from a waterfall, and on them great numbers of fish are constantly jumping. Flocks of seagulls settle on the surface at the entrance of the channel, and after allowing themselves to be carried to its end by its rapid course, fly back to their original position. This not only amused us, but also afforded us a means to gauge in some measure the velocity of the current."

By way of hand signals, the local native people (who were friendly and brought the foreigners fresh salmon) warned them about the dangers of the channels ahead and pointed at the sun to indicate when it would be relatively safe to undertake the passage. But still, the Spaniards nearly came to grief.

They sailed into Arran Rapids at what they expected to be slack water but were carried through "with extraordinary speed." They were swept onward into an adjacent narrows, today called Dent Rapids, site of a whirlpool known as the Devil's Hole. Another whirlpool in the same area, called the Drain, has been likened to a "cosmic washing machine." An entry in one of the Spanish logbooks related that "the continual eddies and whirlpools, now favorable, now contrary, retarded one schooner and advanced the other, rendering it impossible to steer, and carrying us at their mercy." The *Sutil*, "being caught in a strong whirlpool, turned around thrice with such violence that those on board were made giddy." The captain sent out men in a longboat to attach a line to shore, "but at that very moment, caught by another whirlpool, she began to turn around again, and at the first turn tore the cable from the hands of those who were making it fast." With luck and superb seamanship, both ships eventually made it to a safe anchorage.

In modern times, this same area's maze of islands and fjords is home to some of the finest and priciest private fishing lodges in North America. The chance to catch a chinook salmon longer and fatter than a man's thigh is the main attraction, but remoteness and unforgettable scenery are big bonuses. Lofty trees line the shores of tide-torn channels where bears fish and forage, seals and river otters feed, and bald eagles are almost as common as loons or ravens. Craggy snow-capped peaks etch pollution-free skies. There are no roads or ferry service to most of these luxurious fishing outstations, and the steep terrain is so rugged that building airstrips is impractical. Most guests fly in from Vancouver or Seattle in Beaver or Twin Otter floatplanes for several days of personally guided fishing in small boats. Bobbing around among the roaring tidal rapids, rips, and whirlpools adds to the excitement.

I enjoy fishing myself, so I decided to visit several of the lodges, get a close look at the troubled waters, and talk to resort owners, fishing guides, and scientists about the whirlpools and other tidal phenomena. A few of the lodges were essentially men-only places, where middle-aged business types drank, romped nude around the hot tub after the day's fishing, and snapped towels at one another's butts. Most of the resorts, though, were much more refined.

I had read in a magazine about some of the local lore. One legendary fishing guide, Henry Spit, estimated that the Devil's Hole could sometimes reach more than 70 metres (250 feet) across and more than 15 metres (about 50 feet) deep. It sounded to me like typical fishermen's hype. Another very experienced guide, Terry Brimacombe, called this claim "a bit of an exaggeration." In the 1950s his family founded the Big Bay marina and resort on Stuart Island, a rugged chunk of forested rock that makes up one side of the triangle of deep waterways, shallow reefs, and tiny islets that so spooked the Spaniards. The other sides consist of much larger Sonora Island and the mainland shore of British Columbia. The tides flush in and out through narrow channels from three directions, and the timing of the ebb and flood is different in each case, creating a confusion of intersecting and contending tidal streams, but also some great fishing.

Brimacombe grew up on Stuart Island and started guiding in the summers at age thirteen. When he was a child, the area's population of permanent residents was much larger than it is today, with lots of fishing and small logging operations. The kids from these families had to cross tidal channels in rowboats or canoes to get to their one-room schools. Conditions were so hazardous that school hours had to be adjusted each week to allow the children to come and go during the brief periods of slack tide. He survived dicey encounters with all the rapids. "Undoubtedly the worst is the Arrans," he told me, the one where the Spanish ships first entered the maze of tidal madness. "It's really bad. It forms a V in the middle, and on the big tides there's really no way around the whirlpools and boils."

To Brimacombe, these boils were as threatening as the whirlpools. In Arran Rapids, he said, "it does both. There's giant whirlpools and also great upwellings that just frighten you really badly." Anyone negotiating the channel in one of the fast little powerboats the lodges use for fishing has to go around these upwellings, not try to go over them. "If you hit one, you're either going to dive into it or glance off the side and turn upside down, which has happened." To make things worse, the upwellings "appear quite suddenly and can be as much as six feet [two metres] high, more like a dome of water than a standing wave, but this water is moving quite rapidly, and there are also at times quite big waves. And when the water comes up, of course, it's gotta create a hole somewhere," and these are dangerous as well.

Then I spoke to oceanographer Michael Woodward, a fit-looking guy with fair hair and glasses whose job involved placing current meters into the tidal channels. Spending many days each year bouncing around in the currents, he observed the hydrodynamics of their flows and the complexity of the shear forces that were created. It is these contrary energies that generate circulatory vectors and create the whirlpools. "This can happen when you get a main flow and a back eddy along the shore that are opposing each other. There's a fairly well-defined line between the back eddy and the main flow. It might be only half a metre [less than two feet] wide. On one side you might have the main flow running at, say, twelve knots, and on the other side you have a back eddy flowing at two or three knots in the other direction. The shoreline deflects the water and the flow separates."

Woodward estimated that the smaller whirlpools in places like Devil's Hole were around two metres (six or seven feet) deep. I tried to pin down several other guides and scientists and get a realistic estimate of the largest whirlpools on the Inside Passage. The consensus was that these may reach a size as great as over four metres (fifteen feet) deep and twenty-two metres (seventy-five feet) or more in diameter.

But Woodward still thought the main danger was not that a boat would literally be sucked right under, although a person in the water certainly could be, like the dummy in the British experiment at the Corryvreckan. In 1972, a group of American journalists tried to run Arran Rapids in a South Seas–style outrigger canoe. It capsized, and three of the crew were sucked down and drowned. Their bodies were never found. The fourth, a woman, clung to the buoyant wreckage and survived.

"There are some reckless boaters out there," Woodward told me. "I've seen a few myself. A lot of people get into trouble through excessive use of power." Revving up their engines, "they apply far too much power to try to get out of a situation, rather than think[ing]." And thinking means realizing that the boat itself is probably too large and buoyant to be sucked under, unless it flips and fills with water first. His remedy if a boat is swept into one of the whirlpools? "One can cut the power, sit in the middle of the bottom of the boat, hang on, and let the whirlpool spit you out." I asked if that's what happens. "Yes. Almost invariably. Having been spat out a number of times myself," he chuckled. "Usually the boat is large enough that, given the energy in the whirlpool, it collapses."

I expressed some doubt about this, because I knew that the whirlpools in Dent Rapids apparently had enough strength to swing the heavy Spanish ships in circles. But Woodward thought the Spanish vessels were probably *not* being spun around by the power embodied in a single whirlpool. He reckoned that one end of the ship was caught in the main flow of water, which was shearing off in one direction, while the other end was under the influence of a different flow that was coming at it from the other direction. "In Dent Rapids, there's a very strong shear zone on the flood, and along that shear zone there are some quite large whirlpools generated. And those two fast-moving masses of water are going to spin you around like a toy."

Another time, I visited swanky April Point Lodge, a resort on Quadra Island located just south of a tidal choke point called Seymour Narrows. The suites had limited edition landscape prints on the walls and colourful books about birds and fishes on the coffee tables. The dining room offered haute cuisine and dishes like sashimi. The resort had even brought in an expert in *gyotaku*, a Japanese art form and method of creating subtle (and expensive) handmade prints on cloth of the trophy fish caught by the guests.

Seymour Narrows is not far from Arran and Dent Rapids, and has even faster and deadlier tidal currents. At times they run at fifteen or sixteen knots. I wanted to see and feel their effects. Scott Laird, one of April Point's most experienced guides, took me out fishing in a small open outboard runabout. He was a compact, curly-haired guy. Soft spoken and well read, he did not fit the stereotype of the macho outdoorsman at all. "Concentrating on fishing is much like meditating," he told me. "It alters your consciousness." And he preferred catch-and-release fishing, as opposed to killing and keeping the salmon. (Me, I was hoping to bring home a big one to eat.)

We headed out into Discovery Passage, named after Captain George Vancouver's ship HMS *Discovery*, opposite the small city of Campbell River on Vancouver Island. As we approached the narrows Laird made me don a life jacket. Even on this very modest tide, running at perhaps five knots toward us through the passage, it was a minefield of shifting, unpredictable waters. In some places the surface was choppy, in other spots almost glassy smooth, except for the occasional strange, domed upwellings that burst to the surface like depth charges and quickly dissipated. Little whirlpools formed next to us and were swept away by the current.

We joined a flotilla of similar boats that were fishing close in to the green forest wall of Quadra Island, where a back eddy ran counter to the flow of the main tidal stream. We barely got our hooks into the water when the small whirlpools became the least of our concerns. Bearing down on us was a big, bluff-bowed purse seine fishboat that was hugging the shore, taking advantage of the back eddy to buck the current. Its captain was determined to be on his way — time, tide, and our cluster of fishing runabouts be damned. He came on steadily, pushing a high churning bow wave that rolled off to each side. Several of the fishing guides blew their hand-held air horns to warn him, and he answered with blasts of his own much louder horn. At the last minute, when it was apparent that the seiner would run us down rather than stop or change course, Laird gunned our idling outboard and started to move aside. The other small boats also scattererd desperately to get out of the way. But we were a bit too late. The bow wave smashed against us, swept right over the low gunwale, and slammed us like a giant hammer with solid, cold seawater. We were drenched and breathless. Our boat wallowed, half filled with water. The little tape recorder I had with me for interviews was ruined. We bailed quickly as the runabout continued to be buffeted by the waves that followed in the wake of the big fishboat.

Laird was royally pissed off. He cursed the captain and jotted down the name of the seiner as it chugged off into the distance. Later, we filed a joint complaint with the Coast Guard, only to be informed that the seiner's skipper was legally in the right. In a main navigational channel, big vessels enjoy priority. They can plow right on ahead, and the small, more manoeuvrable ones must get out of the way or be run down. Still, bucking the tide looked like recklessness to us. And the region's fishing fleet has often paid the price for taking chances. There's big money to be lost if they don't reach the fishing grounds in time for the official "openings," or if they don't get back quickly to port to sell their catch. So they often take risks by refusing to wait for slack tide, tackling even the worst of the rapids and whirlpools at the wrong time. Laird told me he had seen three large ocean-going seiners roll over and sink in that area. In one case he helped rescue a man, who then died on the way to the hospital.

"It happens nearly every year," Michael Woodward, the oceanographer, confirmed when I checked with him about it. On big tides, powerful turbulence is created downstream from the tightest part of the narrows. These flows are forceful, Woodward told me, because their kinetic energy is proportional to the square of their velocity. In other words, the energy of a mass of water moving at ten knots, as it does at many places on the Inside Passage, is four times as great as at five knots (the speed of peak currents through the Golden Gate, which is considered treacherous because of its powerful tides). At fifteen knots (a speed sometimes exceeded at Seymour Narrows) it is *nine* times as great. This energetic turmoil can create either an upwelling or a deep hole where it breaks the surface. Both can be very steep and appear quite abruptly. So a large fishboat probably cannot be physically sucked under by a whirlpool. But if its stern falls into one of these holes, the vessel can easily be rolled over by the turbulence and capsized. Especially if it is heavily laden with fish.

By comparison, the loss of a tape recorder seemed pretty minor. I could always go to Radio Shack and buy another. I realized that Laird and I had gotten off lightly.

And so it was, too, with the close encounter my friend Daryl and I had had with the whirlpool at Dodd Narrows many years earlier. Although it felt like forever, the whole episode could not have lasted more than five or ten seconds. Daryl's boat made less than one complete turn around the whirlpool before he overcame the frozen disbelief and sprang into action. Adrenalin flowing, he shifted the outboard out of gear and yanked on the starter cord so hard I thought he might tear it loose. The engine sputtered, then caught. The roar of an internal combustion engine had never sounded so sweet. Daryl shifted into forward gear, gunned the motor, and swung it over sharply to point us away from the whirlpool. Fortunately, the old outboard had enough strength to break us free of the tenacious rotating grip.

Out into the widening channel the boat headed as the current carried us quickly away from the narrows. The sun still shone. Seagulls swooped over the water as though nothing had happened. Shaken and sweaty and chastened, we stared at each other and took deep breaths. There was no need even to talk about our escape — that would come later. But we had felt the power of the tide, and how it dwarfed our petty human scale. In my many years of cruising the coast since then, I have never again been careless about timing my approach to a tidal narrows.

The Shape of Tides

The Great Variety of Shorelines Created by Different Tidal Regimes

The Hopewell Rocks were even more stunning in real life than the images I had in my mind from seeing photographs on the Internet. There, spread out below the clifftop park viewpoint with its safety fence, was a group of mammoth, free-standing stone formations as tall as four-storey buildings, all carved and weathered into fantastic shapes. Some of the strange giants, made of a sandstone and pebble conglomerate, had shrubs and small trees growing on their tops. Others formed arches with almost perfectly round openings, or had slots between them that had been eroded into rounded, distinctly keyhole-like shapes. A gentle breeze wafted in off the Bay of Fundy, and small waves lapped on the gravel beach at the base of each megalith. The tide was dropping quickly. My wife, Annie, and I only had to wait an hour before we could descend a set of stairs adjacent to the cliff and walk among the Hopewell Rocks, named for nearby Hopewell Cape.

These remarkable sea stacks along Fundy's New Brunswick coast are a dramatic illustration of how the action of huge tides can physically shape a shoreline. Many have a distinctive hourglass shape, narrower in the middle than at the bottom or the top. This flaring out toward the top, together with the reddish colour of the local sandstone, has also earned them an informal name, the "flowerpot rocks." But it is the narrow mid-section, partway up, that is the real tipoff as to how they are formed.

Photo by Tom Koppel.

The Hopewell Rocks in the Bay of Fundy dramatically illustrate the power of tides to shape the physical world. The erosion caused by the tides has created these hourglass rock formations and has carved out keyholes such as those seen at top (tide falling) and above (close to extreme low tide).

It was easy to picture the tidal cycles acting on them twice a day, year after year over thousands of years. Tides on the Bay of Fundy are semi-diurnal, with two almost equally high periods of high water and two almost equally low periods of low water each day. But not every tide is as high as a spring high tide or as low as a spring low tide. The tidal range during neap tides is considerably smaller. And so, during the neap portion of the fortnightly cycle, the high tides each day only reach partway up along the sea stacks. This means that on every tide each month the sea gets a chance to erode the midsection of each rock, but only during spring tides is the sea able to nibble away at the top and bottom sections. Hence the rocks are more rapidly eroded at the midsections. Some of the hourglass curves and the rounded openings forming archways or keyholes are so smooth and symmetrical that they could almost be graphs in a textbook mapping the length of time the tide gets to act on the rock at each height along a vertical scale. The phenomenon is so odd, and the rocks themselves so impressive, that the Hopewell Rocks are among the most popular tourist attractions in eastern Canada.

Fundy's extremely large tides also produce a very different kind of shoreline. My wife and I were touring the bay to see these huge tides and their effects. The Hopewell Rocks result from currents and wave action impacting on a high and essentially vertical cliff face. But Ice Age geology and post–Ice Age geological processes have left other parts of the coastline along the bay with a very gentle natural slope. A few days after viewing the Hopewell Rocks we stopped the car along the northern shore of the Minas Basin, the northeastern arm of the bay and an area where the tidal range can exceed sixteen metres (fifty-two feet). The tide was low, and I walked out on the nearby beach to the high tide line. This was marked by a gently rising hump or berm that was topped by a thin line of driftwood and some recently deposited dead seaweed. The sea itself was just barely visible as a shimmering line of light in the distance, a kilometre or more away. Between where I stood and that line of water was nothing but a smooth expanse of reddish mud and a few birds pecking around in it.

Unfortunately, it was September, and we had just missed witnessing the spectacular annual migration of 2 to 3 million shorebirds, particularly the semipalmated sandpipers. Following their summer breeding period in the Arctic, they spend a few weeks on Fundy's mud flats each

July and August. Patrolling the edge of the ebbing tide on spindly legs, the birds use their long beaks to dig in the ooze for rice-sized crustaceans called mud shrimp. After doubling in weight, the birds lift off for a non-stop three-day flight over the ocean to their winter home on the northeastern coast of South America. It is only the nutritional bounty afforded by these wide mud flats, which result from millennia of tides carrying in silt and depositing it in thick layers, that allows the birds to follow that life cycle.

The most famous tidal flats in the world are those on the Normandy coast of France near the eleventh-century abbey of Mont Saint-Michel, where the tidal range is almost as large as on the Bay of Fundy, as much as fifteen metres (forty-nine feet). There the sea recedes so far at low tide that often it cannot be seen by someone standing on the flats, which are up to fourteen kilometres (nine miles) wide. When the tide floods in, it can come at the speed of galloping horses. People who do not read the tide tables correctly, or who lose track of the time, don't stand a chance if they have walked too far out onto the flats. Year after year, unwary people are caught and drowned by these killer tides.

Mont Saint-Michel with the tide in.

Mont Saint-Michel with the tide out.

In February 2004, just such a disaster befell a group of Chinese cockle harvesters at another place with very large tides, Morecambe Bay, Lancashire, in northern England. The workers, illegal immigrants working under contract to a Chinese gangster, had apparently not been properly apprised of the danger. As darkness fell, they were overtaken by a rapidly rising tide. Military helicopters with searchlights were called out and managed to save sixteen of them as they floundered in the waves. But at least twenty-one died.

Being overtaken by a flooding tide is one of the quintessential nightmare scenarios for coastal people. The wonderful American mystery writer Aaron Elkins used the Mont Saint-Michel tidal flats as a clever plot device in his 1987 novel *Old Bones*. His protagonist, an anthropologist and amateur detective named Gideon Oliver, keeps a close eye on his watch as he and his wife go out from the abbey for a long and relaxing stroll on the broad expanse of sand and mud. Yet they are still caught out on the flats as the tide roars in on them. As Gideon races for the safety of the distant shore, he realizes that he has been tricked by an evil character in the book, a guy named Ben. Until then, Gideon had not suspected Ben or his motives, but Ben had apparently lied to Gideon hours earlier while reading to him from a French-language pocket tide table. Ben had given Gideon a wrong figure for the time of low-water slack tide and had assured him that there would be plenty of time to hike well out onto the flats and back again. It was a cleverly disguised case of attempted murder. Gideon and his wife just barely escape the insidious rush of the rising waters.

In real life, even the best-informed and most careful people can come to grief because of the tides. While at the Bay of Fundy, Annie and I arranged to meet and discuss the Fundy tides with geologist David Mossman of Mount Allison University in New Brunswick. Mossman is the co-author (with Con Desplanque) of a definitive book-length 2004 study of the Fundy tides. We drove together to the site of rebuilt Fort Beauséjour, a military stronghold from the French colonial era on the bay. There Mossman led us on a hike down to a grassy area atop a man-made dike that overlooked the broad expanse of Cumberland Basin. The tide was low, exposing a mud flat that extended about one kilometre (half a mile) out from the dike. He pointed to some rocks that poked up through the water near the sea's edge and said that they represented the top of a line of glacial till. It was, in other words, a raised ridge of rubble that had been left behind by a retreating glacier at the

Photo by Tom Koppel.

Geologist David Mossman looks out at the tidal flats and water of Cumberland Basin in the Bay of Fundy, with the tide out. The wooden posts in the picture are part of an old dike.

end of the Ice Age. "That's eleven metres [thirty-five feet] below where we're standing," he added.

World sea level, Mossman went on, has been rising incrementally for thousands of years. For reasons that will become clear later in our story, the tidal range on the Bay of Fundy has been getting larger as well. Until about four thousand years ago, that line of glacial till was high enough above the low tide line that trees could grow on it. "It's a fossil forest with four-thousand-year-old wood in it," said Mossman. "You can even burn it. Nice birch, and some big old pine trees. You can see blow-downs there." These were all from trees that grew on the till well after the end of the last glaciation but before high-tide sea level rose as high as it is today. "At a really low tide, you can walk out to there." And that's what led to Mossman's near-death experience.

"A couple of years ago, I helped out a student named James who was doing a dendrochronology [tree-ring dating] project for another professor. One of the assignments was to go out and have a look at that forest. He approached me to go with him and take some saw cuts of the wood," he began. It was around the middle of March. The temperature was about -10° Celsius (14° Fahrenheit), and a brisk wind was coming off the sea. "I knew enough to make sure we were pretty well rigged up for it." Mossman brought along hip waders for the student and wore a wet suit himself under warm winter clothes. They also left some emergency gear on shore in case of wet feet or frostbite.

"We got down here crack on the time of the slack tide," he said. The tide was low, but to reach the fossil wood they had to slosh through water up to their ankles. When they got out there, the student walked around taking precise elevation measurements of the site, while Mossman found a good ancient stump and got busy with a hand-held bow saw. There he was, "beavering away, sawing away at this thing to get a lit-tle cookie," or thin cross-section of the wood. "Well," he went on, warm-ing to his tale as he gazed out to sea from the safety of the dike, "I was sawing away like a fool. And you know, you're supposed keep an eye on the time. You know, slack tide lasts about what? Half an hour at most? And suddenly, *uh-oh*. I got done, and James was finished with his part. And it was getting pretty cold by then." They noticed that the water was already up to their knees and rising rapidly, and a swirling current or eddy had formed that also tended to work against them, pushing them away from the shore and out to sea. "'By golly,' I thought, 'we'd better get our butts out of here.' So we began to walk, and then we more or less ran when we saw how fast the tide was coming in." But running through the rising water was not easy. Meanwhile, the wind had shifted and was now blowing at them strongly from off the land. "That's what really scared me."

Luckily, Mossman had brought along a couple of ski poles to help feel their way while crossing little submerged streambeds, which was a big help, because as they trudged back toward shore the water got deeper and deeper, until it was up to their armpits, and the current kept trying to sweep them out to sea. "James is a smaller guy than I, but fortunately [with the help of the ski pole] he kept his footing. Unfortunately, though, by that time his hip waders had sprung a few serious leaks." So, the student was getting wet and very cold, and the extra weight of water in his boots made it an even tougher slog.

"I tell you, man," said Mossman, grimacing at the memory. "It was like … I wondered to myself, 'Is this the way it's going to be? Huh? With nobody here to see ya? There's no bells ringing or anybody cheering.'" A truly grim and ignominious end. "It was about five in the afternoon and getting dark. It was overcast, with snow blowing in our faces. What a way to go!! And we hadn't even told anybody where we were. I had visions of doom."

"Well, we really got our asses in gear. And we just made it. We just got ashore. James was cold and wet, but we were able to get him into better clothes." And the snow was almost a metre deep, so there was "more fun and games" involved in hiking back up to the car. "Let me tell you. There was a very thin line between the quick and the dead that day."

Coral Reef Flats in the Tropics

Just as large tides shape a shoreline, so can small- to medium-sized tides, although in a number of quite different ways in various parts of the world. And whereas the Hopewell Rocks and wide mud flats of the Bay of Fundy were shaped mainly by purely physical processes, such as erosion and the deposition of layers of sediment, certain other shorelines are shaped primarily by biological processes.

I first realized this a few years ago during a holiday visit to Hawaii, where the tidal range averages well under one metre (only about two to two and a half feet). In fact, most mid-ocean islands have tidal ranges similar to that in Hawaii, or even smaller. A Honolulu friend, Steven Swift, took Annie and me to the most popular snorkelling spot on the island of Oahu. This is Hanauma Bay, where the sea has filled a sunken volcanic crater. The walls of the ancient caldera encircle the sheltered bay except on the southern side, which is open to the Pacific.

Looking out to sea from the beach, most of the first 150 to 200 metres (500 to 650 feet) of the bay is taken up by an inner reef, or reef flat. This is an almost perfectly level platform made up of clumps and clusterings of dead ancient coral, and it is riddled with a maze of channels, crevices, and holes. Growing on top of that old coral like a plastering of cement is a thin crust of much younger multicoloured coralline algae, which is a type of plant that deposits limestone. (Down in the crevices and holes is the occasional growth of young, living coral.) Extremely shallow water covers the reef flat. In most places it is only a few centimetres deep at a normal low tide and is partially exposed when the tide is exceptionally low. At high tide, the water is rarely more than a metre (three feet) deep. Beyond the outer edge of the reef flat, the sea drops off into deeper water where live coral grows in abundance.

The reason Hanauma Bay is popular with snorkellers was obvious. In the deeper channels and crevices in the reef, brilliant life forms thrived in profusion. We saw clouds of colourful fish: yellow and black butterfly fishes, blue striped sergeant fishes, silvery mullets, flamboyantly shaped and coloured Moorish idols, and long, needle-like trumpet fishes. Tucked into nooks in the reef were sea urchins with pencil-like pink and red spines. Lying on the bottom in sandy spots were sea cucumbers, which looked like pale grey sausages.

Best of all, we could enjoy the reef's richness in safety. A large swell was rolling in from the south, probably kicked up by a storm thousands of kilometres away. Waves broke just beyond the fringe of the reef flat, hurling solid water and spume high into the air and generating a roar. But on the inner reef, where we swam over the clusters of fossil coral and algae, we only felt a gentle surging action in the water, a sloshing back and forth in rhythm with the swells. No matter how far out we swam over the reef flat, the water depth remained almost exactly the same. It really *was* flat, almost as if a bulldozer had shaved the top of the reef off at the level of low tide.

I might have thought no more about this, except that when we left Honolulu and flew home I happened to have a window seat. Our plane climbed slowly along the southern coast of Oahu, giving me a perfect view of that entire shore, including Hanauma Bay. I could clearly make out the reef flat there and the rest of the bay, including the telltale colours that indicated water depth. And Hanauma Bay was not unique. Strung out along the southern shore of Oahu was a series of very similar-looking reef flats. Bracketed by rocky headlands, they filled the landward side of every bay, and many of them were more extensive than the flat at Hanauma Bay. I realized that they were a characteristic geological feature in this area, and I wondered how they had developed.

Reading up on coral reef formation, I learned that the development and nature of reef flats is intimately linked to the tides. Because most corals cannot tolerate exposure to the air and full sunlight, except for the briefest periods, the corals grow up toward the sea's surface but remain a bit below low tide sea level. They build on each other to form a reef that is nearly flat along its uppermost layer of growth. Meanwhile, except on the outermost edge, the lower layers die and form solid masses of calcium carbonate, or limestone. If this process takes place along a steep volcanic slope, as in Hawaii, there is a limit to how far out from the island the reef can extend, because most corals cannot live at depths greater than about 45 metres (150 feet).

Another limitation on the growth of corals is wave action, which can break off the more fragile branches of certain corals and scour others with sediment or the detritus from broken coral. This means that most corals on exposed shorelines can normally only grow up to within about two or three metres (six to ten feet) of low tide sea level. Even at that depth, they can be severely damaged by the waves from extreme but infrequent events such as hurricanes or tsunamis, and the shallower parts of the reef often take decades to recover.

However, where tides are small, as in most of the tropical and subtropical Pacific and Indian Oceans, coral reefs develop an additional feature known as a reef crest. This is a raised ridge along the outermost edge of the reef flat that breaks the surface at low tide and is usually just below the surface at high tide. The reef crest is made up largely of the debris from coral that had been growing in slightly deeper water along the outer slope of the reef and was thrown up onto the reef's edge by wave action. There it gets cemented together by coralline algae. The reef crest takes the brunt of the battering from energetic waves. In behind it, to the landward side, the reef flat tends to be a bit deeper. It generally fills in with pulverized coral and other detritus that smothers most living coral.

Reef flats are common throughout the tropical Pacific and Indian Oceans, especially where reefs fringe volcanic islands. They also exist along mainland shores in places like the Gulf of Aqaba in Jordan. In most of these areas, the tidal range is very small, which allows them to grow right up to low tide sea level. There are some places, however, where coral reefs grow in the presence of a larger tidal range. Madagascar, with a spring tidal range of three to four metres (ten to fifteen feet), is a prime example. There, the reef flats have a slightly different character. The reef crest tends to be a considerably higher ridge, which it must be to break the force of energetic waves if they happen to come at a time of much higher tides. The ridge is often made up of large blocks of debris that have broken off from the outer coral growth and have been hurled up on shore over the years by the rare giant storm or tsunami wave. But even this higher reef crest is not sufficient to keep the sea from surging over the reef flat. Wave action carves widely spaced tidal channels through the crest, so at high tide there is often a much greater depth of water over the reef flat than in a place like Hawaii. At low tide, however, the reef flat is often exposed.

Boulder Barricades, Salt Flats, Beach Rock, and Oyster Reefs

An extreme contrast to the warm tropical seas where coral grows is the situation where cold seas freeze over in winter. Among the world's least known shoreline features are called boulder barricades, which are created by the combined action of ice and tides. These are strange-looking but distinctive lines of large rocks that can run for many kilometres, following the low tide line on certain subarctic and arctic shorelines, including parts of Scandinavia and Alaska.

The best studied area with boulder barricades is Canada's remote region of Labrador, and in particular Makkovik Bay, a subarctic fjord with a gently sloping shoreline and tides in the high microtidal to low mesotidal range. (The mean tidal range for all tides is 1.4 metres, or 4.6 feet, while the mean tidal range for spring tides is 2.2 metres, or 7.2 feet.) Each winter solid sea ice forms in the bay. Eventually tight pack ice extends as much as 100 kilometres (60 miles) offshore, which breaks the sea swell and creates a very low energy shoreline within the bay. The solid, floating ice close to shore gradually attains a thickness of about 2 metres (6.5 feet), and it rises and falls regularly with the tide.

Courtesy of Jim Dempsey, Inco Ltd.

Labrador's boulder barricades.

As the ice freezes in early winter, it does so from the bottom of the first layers of ice downward. That is, new ice is created day by day on the underside of the older ice. In deep water, this simply creates an increasingly thick floating ice sheet. But close to shore the situation is different. Where the broad intertidal flats slope very gradually, the lower surface of the already formed ice often rests on the bottom at low tide. Bottom sediment, such as gravel, routinely becomes frozen into the lowest layer of the ice.

Some of the intertidal flats, however, are also strewn with sizeable boulders, left there by the last glaciation. When the relatively thin ice of early winter rests on these boulders at low tide, its weight creates a distortion. The boulders push their way up through the ice, creating conical bulges that show on the surface and are locally called ballycatters. When high tide follows, the ice sheet floats and rises off the boulder, but because of the low energy situation, the sheet does not move laterally to a significant degree. This means that during the next low tide, the ice sheet settles back onto the same boulders. The boulders again push their way up into the weak points or breaks in the ice that the ballycatters represent. As the ice thickens by "freezing down" over the ensuing weeks, the boulders become frozen solidly into the bottom layer of ice. And they get lifted off the bottom by the ice on high tides.

Then comes spring. By June, the ice on the bay begins to break up. Cakes of ice, or floes, some quite large and holding boulders up to two metres long, float freely. Winds and tidal currents move them around the bay, often transporting the encased boulders at least several kilometres from their source. Eventually, the floes run aground at the edge of a tidal flat. Because the average thickness of the ice is about the same as the mean tidal range, they tend to ground out precisely at the low tide line. There the ice eventually melts, leaving the boulders clustered along that tide line. It looks as if some Ice Age giant had intentionally laid them down there as a stone wall.

Another less familiar coastal feature created in part by the rise and fall of the tide is the coastal salt flat, which is commonly known in the Middle East by its Arabic name, *sabkha*. There are similar extensive salt flats (also called salt barrens) on the coast of Baja California, Mexico; on the Pacific coast of Ecuador; at King Sound in northwestern Australia; and on the west coast of Panama, where they are called *albinas*.

Coastal salt flats are usually devoid of plant life beyond microscopic size, because the salt content of the soil is so high. They form in the upper part of the intertidal zone on very low gradient shorelines that are subjected to a combination of several distinct conditions: extremely dry air, usually from a desert; intense desert heat; strong winds; and sizeable, usually meso- to macrotidal, tides. Each time there is a high spring tide, rapid evaporation leads to the deposition of the salt that has been carried onto shore by the sea water. Often desert winds blow the salt around as well, carrying it well inland. Those winds also bring to the coastal salt flats a mix of fine-grained mud and fine sand, which is then deposited there. Since the tides generate a periodic immersion of these sediments, a type of blue-green algae often grows on them. Rapid evaporation and intense heat then binds the material together to form thick mats. The salt flats can be very extensive. In many places on the Arabian Coast of the United Arab Emirates the intertidal zones are one to two kilometres (up to a mile or more) wide. A series of *sabkhas* there stretches for some 320 kilometres (200 miles).

Elsewhere, conditions in some intertidal zones lead to the formation of a hard deposit called beach rock. In tropical climates, dissolved calcium carbonate acts as a natural cement that binds particles of beach sand and other sediments (along with bits of shell, coral, and other materials) into a solid form of sedimentary rock. The calcium carbonate is mainly derived from the remains of zoo plankton in the ocean water. Each tide carries some of this dissolved chemical compound into the intertidal zone, where it leaches into the beach materials. Then, after the tide goes out, the intense sun and high temperature go to work, evaporating the water and baking the residual solids.

It usually forms into flat slabs that look a bit like flagstones, and there is often a downward dip to seaward. Beach rock can be very old and full of beach fossils. Since it only forms in the intertidal zone, if found in situ it is often an excellent indicator of ancient (and changing) sea level, especially where the tidal range is very small. Where changing sea level and erosion have undercut layers of beach rock, it breaks loose. In many places that lack ordinary rock (such as on coral atolls), it is a convenient building material.

But not all beach rock is old. Unlike some other sedimentary rocks, it can form quite rapidly. The tourist board of Vanuatu, in the southwestern Pacific, points to some areas of beach rock there as a historical curiosity: "In places like Espiritu Santo this beach rock has welded not only sand, but the remains of WWII machinery dumped at places like Million Dollar Point, into one huge long beach of naturally cemented war refuse. It's an amazing sight for not only war machinery was dumped here. Million Dollar Point may well have the distinction of being the only beach in the world made from rock embedded with tens of thousands of Coca Cola and 7-Up bottles."

Another shoreline feature that reflects highly specific tidal conditions is the oyster reef. Where the American or eastern oyster (*Crassostrea virginica*) grows in profusion, such as in certain estuaries on the South Atlantic Bight (between Cape Fear, North Carolina, and Cape Canaveral, Florida), they can build up over centuries to form large reefs covering hundreds of hectares and up to one and a half metres (five feet) thick. As a U.S. Fish and Wildlife Service study states, "some reefs are many kilometers in length." Tiny free-floating larval oysters are attracted by a protein on the surface of existing oyster shells. They settle and attach themselves to older living oysters and grow on top of them. The older oysters eventually die, but by then a new layer of live oysters has established itself as a colony above the old ones. The resulting reef is a huge buildup of almost solid calcium carbonate.

Individual oysters can live in deeper water, but dense colonies of oysters are characteristic of the middle range of the intertidal zone. The base of the oyster reef is usually a little above the elevation of mean low tide, and it grows upward from there. Oysters can stand regular exposure to air and bright sunlight, and as filter feeders, they thrive best where the coming and going of the tide brings them a steady supply of microscopic food. Not surprisingly, therefore, the upper limit of an oyster reef is determined by mean high tide, with the living oysters growing up to about half a metre (one and a half feet) below that level. The largest mean tidal range along the South Atlantic Bight, about 2.2 metres (7.2 feet), occurs around Sapelo Island, Georgia, and the most impressive oyster reefs are found there as well. Like coral reef flats, they are almost perfectly level on the top surface and look gray when exposed and dry, but greenish-brown when wet, because of a thin film of algae that lives on the shells.

Deltas, Estuaries, and Shifting Beaches

Boulder barricades, *sabkhas*, beach rock, and oyster reefs are localized and unusual shoreline features created in special situations. There are also far more common tide-related features that are found in many parts of the world. For example, coastal geomorphologists (scientists who compare and analyze the configurations and dynamics of various shorelines) have noticed that some coasts are wave-dominated while others are tide-dominated. Both waves and tides are marine processes that transport loose sediments such as sand, and in doing so sculpt beaches and river estuaries, but they act in somewhat different ways.

Wave energy impacting a shoreline can be very powerful. Often this is due to a region's strong winds and the exposure of the shoreline to the open sea and its long-distance swells. If the tidal range is relatively small or negligible, the waves dominate. In places such as the Great Lakes, where the tide is barely measurable, waves and wave-driven currents can shape beaches entirely on their own. Acting in combination with the outward flow of water at a river mouth, waves can create the distinctive features of an estuary. The beaches and estuaries on wave-dominated shorelines are quite different, however, from those where the tides are large and the energy in tidal currents is dominant over wave energy.

One typical kind of wave-dominated shoreline develops where powerful wave action carries sediments in toward shore from the shallow offshore seabed, creating sandbars or long, narrow barrier islands that run parallel to the coast. This occurs along much of the eastern shore of the United States. Part of the New Jersey coast, where Ocean City is built on a barrier island, is a good example. So are the Outer Banks of North Carolina, where Cape Hatteras is the best-known area. Along that coast, tides are generally only in the upper microtidal or lower mesotidal range, rising and falling less than two metres. But the wave action off the often stormy North Atlantic is very strong.

Where a river comes down to the sea along a relatively straight and open coast, the nature of the resulting shoreline also depends largely on the relative power of waves versus tides. River sediments carried by the flow of fresh water will tend to fan out onto the adjacent continental shelf and form a delta. Deltas usually consist of raised sand ridges, called bars, and submerged ridges, called shoals, with channels between them where water can flow. But the shape and orientation of these features is related to the relative strength of the waves and tides. Where tides are small and waves are moderate to strong in power (such as on an open coast exposed to the ocean swell), those river sediments will tend to form bars and shoals that sweep in an arc roughly parallel to the coast. (See Figure 5.1 a.) But if the tides are large and the wave energy is relatively weak (as on a more sheltered coast), a series of long, narrow bars and shoals will form that are aligned with the river's flow as it fans out into the ocean. (See Figure 5.1 b.)

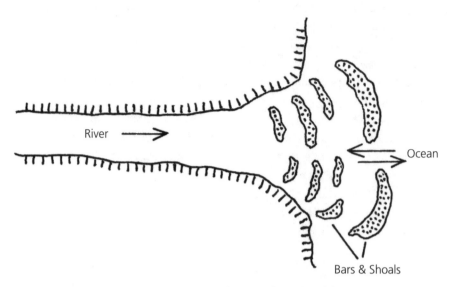

Figure 5.1 a: A wave-dominated delta.

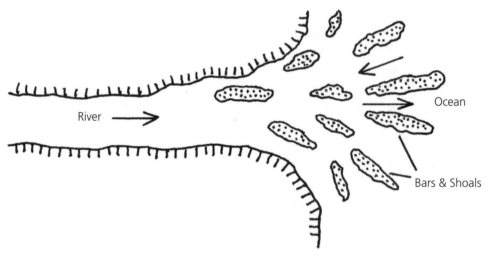

Figure 5.1 b: A tide-dominated delta.

A somewhat similar set of features usually forms where a river meets the sea not along an open stretch of coast but within the confines of a deeply indented estuary. (Such estuaries exist mainly where, during glacial times, a valley came down to the sea. That valley was subsequently drowned when world sea level rose 100 metres [325 feet] or more follow-ing the last glaciation.) If the waves at the mouth of such an estuary dom-inate over the tides, a barrier sandbar, which is exposed even at high tide, will normally form, partially blocking the mouth of the estuary. There will always be a narrow channel to let the river's water flow out past the barrier, but on the flood tide ocean water will only flow into the estuary a relatively short distance and without great speed. (See Figure 5.2 a.)

An estuary will look quite different, however, in a coastal region with a large tidal range, especially if the waves are weak. In that case, strong tidal currents will sweep the sea water far into the estuary and partway up the river on a flood tide, counteracting the river's flow. Then, on the ebb, the combined falling tide and river flow will sweep rapidly back down the river channel and out to sea. Repeated with each tidal cycle, this flow of water carries sediments along with it and, over time, deposits them as sandbars that are aligned with the river's flow and are exposed at low tide. Often there are multiple tidal channels between the bars. Estuaries in places with large tidal ranges also tend to be relatively long and narrow. (See Figure 5.2 b.) In temperate climates, there are usu-ally salt marshes lining the banks of the main river channel. In tropical or subtropical regions, mangrove often grows in those same areas. By contrast, wave-dominated estuaries are generally shorter, and they often have a broad but shallow and muddy inner basin behind the barrier across the entrance.

The shape and nature of beaches also depends in large part on the relative strength and size of the waves and tides. (Although other factors come into play as well, including the underlying gradient of the land just offshore and the amount and type of material, such as sand, that is avail-able to be recycled by waves and tidal currents in the processes of beach formation.) How waves act as they sweep in onto a beach is surprisingly complex, and beaches respond to the wave action in dynamic ways. The same beach, for example, can have an entirely different slope and appear-ance in winter, when it is pounded by powerful storm waves, than in summer, when it is subjected to more gentle wave action. Beach forma-tion is an ever-changing but eternally repeated process that can either erode a shore or build it up, depending on circumstances.

Marine scientists distinguish between different kinds of wave action on beaches. Some waves break or plunge dramatically in a cascade of

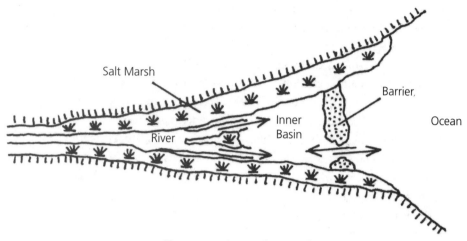

Figure 5.2 a: A wave-dominated estuary.

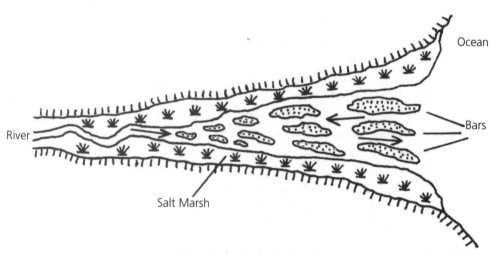

Figure 5.2 b: A tide-dominated estuary.

foam as they bottom out in shallow water. Others peak more gently, and their crest only spills onto the beach. Another type just surges up along the slope of the beach, without really breaking at all, and then washes back down again. Each kind of wave is different in how it transports beach materials, sometimes building up the beach in one area while eroding it in another. Waves (acting in conjunction with currents) can change the slope of a beach, or they can transport sediments laterally along the beachfront. In the absence of sizeable tides, waves and currents generate a dynamic set of alternately creative and destructive processes that often find a kind of balance or equilibrium. The beach looks roughly the same day after day, except for the change of seasons between summer and winter.

Where the tidal range is large, however, these wave actions and their resulting effects on beach materials — all these processes — get shifted across the beach profile once or twice a day. They move up the beach slope toward shore and then down again toward the sea, in lockstep with the flood and ebb. And they impact the beach somewhat differently at each stage. Take one typical kind of beach in an area with large tides and a gently sloping underlying landform. The back of the beach tends to be narrow and quite steeply sloped. At high water, waves may break with great force close to, and directly against, that slope. These powerful waves can carry quite coarse beach materials, including gravel and pebbles, with them. They hurl these materials against the back of the beach, which builds it up. This is what makes it steep. They also tend to gouge it out, sculpting it into a concave shape.

To seaward of that zone, however, the gradient of the beach becomes much gentler. When the tide is out, much of this gently sloping surface is exposed, and the same gradient continues out to sea under the water in the subtidal zone. When waves sweep in at low tide over the shallow subtidal area, they dissipate their energy with much less foam and fury over this very extensive zone. Only much finer beach material is carried and deposited. A smooth, fine-grained, and much more gradually sloping low tide zone results.

There are so many variables, including the local tidal range, at work on different beaches around the world that it would be tedious to catalogue all the resulting configurations. Entire technical books have been written on this subject. In some cases, though, the shape of the same beach changes over the course of a single fortnightly tidal cycle.

In the 1990s, geographers Gerhard Masselink and Bruce Hegge conducted a study of how waves, currents, and tides affected two beaches along the central coast of Queensland, Australia, during periods of more than two weeks. Both neap and spring tide conditions were observed, as well as the daily changes from low water to high and back down again. The beach profiles were measured daily by surveying along transect lines running from the highest part of the beach down into the lowest intertidal zone. Sediment samples were collected at regular intervals and studied to see how quickly the particles settled in tubes of water. Flow meters measured the speed and direction of currents just off the beach. Combining the data gave a complete picture of the changing beach and shore conditions over the span of a fortnight.

Both beaches had tides that were in the mesotidal range (two to three metres, or six to ten feet) during neap tides but reached well into the macrotidal range (four to six metres, or fifteen to twenty feet) at large spring tides. The study found that the tides, although not the only factor at work, played a major active role in forming, sculpting, and reconfiguring the beaches.

One beach, which they designated a low tide terrace type, had a profile like the one shown in Figure 5.3. There was a relatively steep gradient through most of the vertical range from where mean high water lapped on the beach during a spring tide down to the level of mean low water during a neap tide. Below that elevation the slope became much gentler

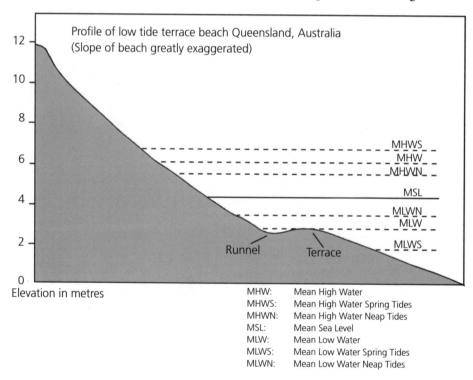

Figure 5.3: Profile of the low tide terrace beach in Queensland, Australia.

and there was a terrace of fine sand that was exposed at extreme low water during spring tides. This terrace sloped off into the subtidal zone. In the course of a tidal cycle (from neap to spring and back to neap again) this beach went through "subtle, but significant changes" involving "minor, but consistent adjustments of the intertidal profile" in response to the changing tidal ranges. Changes to the low tide terrace were almost negligible. But higher up along the intertidal slope, as the tides got larger each day (going from neap to spring tides) the waves eroded away the upper portion of the slope, scooping it out just a bit. At the same time, the lower part of the beach profile, just above the terrace, accreted. In other words, sediments were deposited there, building it up slightly. The overall result was a flattening of the beach profile, leaving a slightly less sloped gradient down to where the slope met the terrace. But then, a week or so later, as the tides began to get smaller again, the process reversed itself. "The upper intertidal profile accreted ... whereas the lower profile underwent erosion," so the overall beach profile steepened once again.

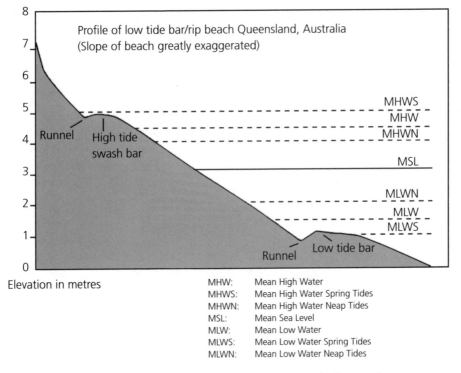

Figure 5.4: Profile of the low tide bar/rip beach in Queensland, Australia.

The second beach Masselink and Hegge studied, which was subjected to a somewhat smaller tidal range, was what they call the low tide bar/rip type. (See Figure 5.4.) It is characterized by what is known as a swash bar, created by sediments that rush up the beach as waves collapse a bit lower along the profile. That bar, which is exposed during neap high tides, runs parallel to the high tide line and is situated well up in the intertidal zone. There is also another bar in the lowest part of the intertidal zone that is only exposed between waves during the lowest of spring tides. (On a spring tide, the lows are lower than on a neap tide, just as the highs are higher.) To the landward side of each bar the beach profile dips slightly, leaving a trough or runnel. At high spring tides, water can run along the upper trough parallel to the beach, and at low spring tides it can run through the lower trough. (The water running parallel to the beach takes the form of rip currents that redistribute sediments.)

The scientists also found that at neap high tides a swash bar formed lower down on the beach profile, and it, too, had a trough or runnel to landward of it. Then, as the tidal range increased from neap to spring tides, sediments moved and the neap high tide swash bar disappeared. But when spring tides had been at their peak for several days, sediments were reshuffled across the beach again. First the high tide swash bar disappeared, and a few days later the low tide bar vanished as well, leaving a relatively uniform gradient for the last few days of each cycle. As with the first beach, the profile of the second beach also changed over the course of the fortnightly tidal cycle. On both beaches, these changes recurred roughly twice a month. The tides shaped and reshaped the shoreline in a dynamic and eternally repeated process.

Age of Discovery

European Explorers Learn How Complex the Tides Can Be

Back in the 1980s, I went for a cruise along the British Columbia coast with my friend Philip Newton on his small sailboat *Loon*. Although we had a good chart, we did not study it carefully enough as we approached an anchorage on Gabriola Island to spend the night. There was almost no wind, so we took down the sails and motored in. Then we made the dumb mistake of judging the shape of the bay by its visible exposed shoreline. So we headed straight for the middle of the bay, where we intended to drop the hook.

We had failed to notice, however, that the underwater profile of the bay — the way it would look at low tide — was quite different from the contour that we could see at what was then nearly high tide. A shallow shelf of rock jutted from one side well out into the bay. It was clearly shown in the scattered numbers on the chart that represented depth at low tide. But, lulled by the calm conditions and soft light of late afternoon, we blithely motored in over that shelf.

Suddenly, our keel struck bottom. Dumbfounded, we gaped at each other in horror as our slow forward motion instantly ceased. Realizing what had happened, Philip quickly threw the engine into reverse. The prop revved and protested, but *Loon* did not move. We had run aground. Stuck solid. It was every yacht owner's worst nightmare. Fortunately, with little wind, only tiny waves rolled into the bay. We rocked up and down a bit, but we could feel only the slightest grinding vibrations. Trying to swallow our panic, we checked the tide table. And what a relief! It told us that the tide was flooding and would continue to do so for at least the next hour. That afforded a measure of hope.

We lurched into action, inflating the rubber dinghy and putting down an anchor, with the thought that it might help in pulling ourselves off the rocks. After that, all we could do was wait for time and tide. As dusk fell, the boat's motion became livelier. We were floating. Philip fired up the engine and backed *Loon* off the underwater obstacle. Guided this time by the chart, we motored around the line of hidden rocks and anchored in deeper water. Then we broke out the scotch and settled in for the night. In the morning, we headed off, personally humbled yet with no significant damage to the boat.

Not all vessels have gotten away so lightly after running aground in an area with large tides.

Carefully Measuring the Sea's Rise and Fall

During the Age of Discovery, as Western mariners left their more familiar European waters and explored the far-flung shores of the world's little-known oceans, they had neither good charts nor books of tide tables to keep them out of trouble. Instead, they gradually gained practical experience of tidal phenomena along distant coasts the hard way, through careful observation and risky trial and error. In many cases, they ran aground on rocks and reefs, just as my friend and I had done.

Take Captain James Cook's ship *Endeavour*, for example. It suffered nearly fatal damage when, in darkness, it ran onto a coral reef off the coast of Queensland, Australia, in 1770, during Cook's first great voyage of discovery. Then the tide ebbed, leaving *Endeavour* leaking badly and stuck in a desperate situation. To lighten the ship, the crew jettisoned much of their drinking water, ballast, even six of their heavy guns. When the next flood tide came, it was not quite high enough to float the ship off, even though they had rid themselves of fifty tons of weight. Then the tide ebbed again. As the gripping official account relates, "Their only hope now depended on the midnight tide," the next flood. Luckily for Cook and his men, there was a sizeable diurnal inequality in the tides on this part of the Australian coast. The second high tide proved to be higher than the disappointing first one. As the tide began to rise, "the leak likewise increased to such a degree, that three pumps were kept going." At last, however, "the ship floated, and was heaved into deep water." By then, there was over one metre (nearly four feet) of water in the bilges, and the crew was exhausted from the pumping. They headed straight for the Australian mainland to make repairs.

Similarly, Captain George Vancouver's ships *Discovery* and *Chatham* both ran onto hidden rocks within hours of each other while surveying off the northern end of Vancouver Island in 1792. As with Cook's voyage, the crews took drastic action, such as hauling down the topmasts, shoring and propping up the ships, and lightening them by jettisoning excess water and ballast. In both instances, a rising tide eventually floated the ships off. In all three cases, it was extremely fortunate that the ships had not run aground while the tide was at or near its very highest level — or when a heavy swell was running, which could have pounded them to pieces.

Running onto rocks and reefs was one of the routine risks that mariners faced on voyages of exploration. The main way of avoiding hidden dangers was to sound for shallow areas or obstructions with the lead line, often from a small boat that went out ahead of the ship. Any hazards were then noted and recorded in logs and journals and on future charts. But learning about the local tides was crucial as well. In each new area explored, ships' officers observed how much the tides rose and fell, how swiftly the tidal currents ran, and the direction of these currents on ebb and flood. This information, which supplemented the journals and charts, could be invaluable for the next captain who sailed that same coast. In an era when sailing ships depended entirely on fickle winds, an unanticipated tide, especially one that generated a strong current, might easily sweep a ship to destruction on a reef or rocky shore, even if the captain knew quite precisely where such a danger lurked. Gaining a clear understanding of a region's tidal regime was almost as important as charting the rocks and shoals.

It was an enterprise that stretched over hundreds of years. What gradually emerged was something resembling a patchwork quilt, a worldwide picture of the tides that was far more complex and diverse than Europeans had initially imagined. The accumulation of empirical data on tides generally preceded advances in tidal theory, which is not surprising, since the point of theory was to explain how things worked in the real world. But the very complexity of the tides, as revealed through practical observation and measurement, made them remarkably difficult to account for. This forced tidal theorists to come up with a succession of new ideas, hypotheses, and refinements of theory over the ensuing centuries.

By the mid-seventeenth century, European ships had explored large parts of the North and South Atlantic and most of the Pacific coast of America. They had crossed the southern Pacific and established depots on some of its islands, sailed the shores of the Indian Ocean and around southern China, and even rounded parts of Australia. Many of these voyages encountered surprising or dramatic tides. They brought back at least rudimentary charts and records of what they encountered to their home countries, including information on the tides. But this was by no means a coherent or reliable body of tidal knowledge.

For one thing, the European powers were in commercial competition (and much of the time actually at war) with each other. Navigational knowledge was power, so they had no interest in disseminating what their captains learned during their voyages. Maps and ships' logs and journals were often treated as state secrets.

Nor is the vertical rise and fall of the tide on a shore as easy to measure accurately as it might seem. Ocean swells surging up and down along an open beach or cliff make it difficult to determine the exact average level of the sea at any given time. Often the best that could be done was to note the apparent height of the water on dock pilings or fixed poles within a relatively sheltered harbour. Even in such places, however, waves could distort the readings, especially in windy weather. The first simple tide gauges capable of eliminating such effects were not invented until the 1830s. And someone still had to be there to note the level of the tide and the exact time on a regular basis. Not until much more sophisticated, self-registering tide gauges were installed in major ports (in the mid to late nineteenth century) was it possible to obtain truly accurate and reliable records of the rise and fall of tides over long periods of time.

Courtesy NOAA Photo Library.

A scientist checks a self-registering tide gauge.

In short, although much information flowed back to Europe about the behaviour of tides in foreign waters, it was spotty, of dubious accuracy, and not widely reported. A partial exception was the early British, French, and Dutch colonies along the northeast coast of America, where it became apparent that the tides roughly resembled those of Britain and France. That is, the dominant pattern was of semi-diurnal tides that were quite regular, with each high or low tide being very nearly the same size as the one before or after it. This initially reinforced the notion (which proved to be quite false) that such a tidal regime was the norm for the world's oceans.

It was not until the second half of the seventeenth century that subtleties gradually emerged in the overall picture of how tides worked. These resulted mainly from systematic observations and measurements carried out along the coasts of Britain and France. The first initiatives came in England at the Royal Society of London, which was founded in 1660 and held regular meetings at Gresham College. Beginning in 1665, letters and papers on the tides were frequently received and published in the Society's *Philosophical Transactions*. This was well before the first major theoretical breakthrough, Newton's equilibrium theory of tides. At the time, Galileo's and Descartes's concepts of tides, which did not involve gravity, were still prevalent.

The Reverend Joshua Childrey was an archdeacon of Salisbury Cathedral and an enthusiastic astrologer who lived near Weymouth on the coast of Dorset. He reported to the Society on continual observations of weather and other phenomena that he had made in the 1650s, and supplemented this with information from even earlier historical records. He had noticed certain instances of unusually high tides and pointed out that such variations seemed to coincide with times when the Moon was near perigee, that is, at the point in its orbit closest to the Earth. David Cartwright, today's leading British tidal scientist, calls Childrey "the most interesting of [the] early amateur observers and collectors of data" on the tides, and adds, "We now know ... that the closeness of the moon to perigee or apogee has a stronger effect on the amplitude of spring tides than the passage of the equinoxes," that is, when the Sun is directly over the equator.

Another tidal pioneer was Sir Robert Moray, an influential Freemason and one of the founding members of the Royal Society. Moray submitted a paper for the first volume of the *Philosophical Transactions* about unusual tides he had observed over a month's time in the Outer Hebrides of Scotland. He was the first person to perceive a twice-monthly progression in the size of the tides from spring to neap and back again to spring, and he implied that this pattern related to the phases of the Moon. As Society minutes noted in 1666, Moray saw an "increase of the tides from the quarter [i.e. at quadrature, when the Moon and Sun are at right angles to each other relative to the Earth] to the spring-tide [when the Moon and Sun are in alignment with the Earth], and their decrease from the spring-tide to the quarter." And he suggested that "these increases and decreases, these risings and fallings" might follow a curve like "that of sines, or something near it." In other words, when each day's high and low tides are drawn out on a graph over the course of a month or so, they might trace out the graceful and familiar shape of a typical sine curve. There would be no abrupt or irregular rises or falls, but a smooth progression. A glimpse at a modern tide table reveals that, in some places, tidal highs and lows do roughly follow such a curve. It was just the kind of observation that Newton's theory, which was not to appear for another twenty-one years, went a long way toward explaining.

Also in 1666, John Wallis, professor of geometry at Oxford, sent the Society an essay on "the Flux and Reflux of the Sea." It basically supported, but also elaborated upon, Galileo's idea that tides were caused by changing accelerations and retardations of the water at different points on Earth as our planet rotated on its axis and moved in its orbit. One Royal Society member, Dr. Jonathan Goddard, objected that if Wallis's explanation of the tides were correct, there should be spring tides only once a month, not twice, as was clearly the case. Other members pointed out that around the coasts of Britain spring tides usually occurred two or three days after the new or full moon, but that there were reports of places in the East Indies where spring tides occurred when the Moon was in quadrature.

Wallis responded to these objections and discrepancies by admitting that reliable tidal information was lacking and proposing that, in order to help prove or disprove his theory of tides, the Society should arrange for observations of spring tides to be made over a number of months. This stimulated a number of initiatives and amounted to the first organized and clearly targeted study of the tides. One member offered to contact people along the Thames to help in the effort. Sir Robert Moray proposed taking regular measurements of the tide's height in a place with a large tidal range, such as on the Severn River and Bristol Channel, between Wales and England, where the tidal range is as much as 15.4 metres (over 50 feet). He wanted observers to keep records of the rise and fall for months, possibly years, and also to measure the speed of tidal currents by throwing out a log on a line, as was done on ships at that time. A key goal was to obtain an accurate picture of how the tides rose and fell both during each twice-daily tidal cycle and over the course of the fortnightly progression from neaps to springs and back again. This should reveal whether they followed a sine curve or not. Wallis's elaborate plan was never realized, but the Royal Society did inspire some correspondents to make (and report back) observations from a number of places in Britain and from as far away as Bermuda and Portugal.

One respondent, Samuel Colepresse, reported on measurements of tides at Plymouth in southern England, pointing out that the largest spring tides came on the third tide after a new or full moon (the syzygies), and that these highest tides were followed by the lowest ebbs. A report from Bermuda claimed that the tidal range there was never more than five feet (about one and a half metres). (Actually, according to modern measurements its maximum is about one metre, or three feet.) And from North America came information about the huge Bay of Fundy tides, which had apparently been the subject of contradictory reports. John Winthrop wrote that he sought further information, but "it is certaine that the water floweth and ebbeth much in that sea, above all the other places of these parts, that I heare of." Moray urged Winthrop to provide more detailed information and expressed his doubts about the Bay of Fundy: "I can hardly think the Ebbes & floods can be greater there than on the Coast of France where the tide rises 14 fathom upright: and you cannot but know that in the [Severn River of southwestern England] it flows Ten Fathoms. In a word, write what you hear, & enquire further."

Perhaps because proper tide gauges were non-existent, these early claims regarding tidal ranges were greatly exaggerated. Fourteen fathoms would be almost 26 metres (84 feet), whereas in fact tides on the coast of France do not exceed about 15 metres (49 feet). Ten fathoms would be 18 metres (60 feet), but the tidal range on the Severn is "only," as mentioned, a maximum of 15.4 metres (50.5 feet). And there was an additional complication, namely establishing just *when* the tide has begun to turn. Slack water usually lasts at least a half-hour or so, after which the tide rises or falls quite slowly at first. When Britain's Astronomer Royal, John Flamsteed, produced a tide prediction table for the Thames River in 1683, he bemoaned "how difficult it is to determine the Time of an High-Water exactly."

One thing that the Royal Society's respondents seemed to agree upon was the prevalence in European waters of semi-diurnal tides with near equality. The observation was by no means new. In ancient times, Poseidonius had observed regular twice-daily tides on the Atlantic coast of Spain. Julius Caesar may not have had a good sense of the timing and size of the tides on the coasts of France and Britain, but he would have seen them ebb and flow twice a day. Likewise, the Venerable Bede observed two approximately equal high tides and two low ones each day on the coast of northern Britain. And when Europeans first explored and established colonies on the coast of North America, they found a roughly similar tidal regime on the western side of the Atlantic. The tidal range generally increased as one proceeded north from the Carolinas and Virginia to Massachusetts and Nova Scotia, and the overall pattern was one of regular semi-diurnal tides. If it was high water now, there would be another high tide of roughly equal height in a little over twelve hours. These high tides would increase or decrease in height only very gradually, day by day, in the course of the cycle from neap to spring and back again. And the currents generated by flooding or ebbing tides generally also reversed themselves predictably twice a day.

It was quite a surprise, therefore, when Europeans first learned that in many other parts of the world the tides behaved differently. This information filtered back to Europe and the American colonies in bits and pieces. At first, these other tidal regimes were thought to be quirky and anomalous, exceptions to the supposed norm of twice-daily tides. Even today, in many North American geography books or school texts semi-diurnal tides of approximately equal height are the only ones discussed. Where tidal theory is concerned, both children and adults read a drastically simplified account of Newton's two tidal "bulges" sweeping around the world like clockwork, bringing with them two high tides and two lows each day. Yet on the entire west coast of the United States and Canada, the main pattern is one of mixed tides. There are two highs and lows a day, but for much of each month, one of those highs or lows is quite different from the other. And on the Gulf of Mexico there is generally only one high and one low each day. Not until the early twentieth century did tidal theory come close to accounting for these differences.

An Odd Phenomenon in Indo-China

The first reliable report of strangely behaving tides on distant shores reached the Royal Society in 1678. Francis Davenport, an Englishman and an employee of the East India Company, carefully observed the tides at a place called Batsha on the Gulf of Tonkin (located on the Do Son Peninsula along the northern coast of what is today Vietnam). His purpose was to help the company's large and deep-draught ships safely cross a sandbar that blocked the estuary of the Red River. What he noticed was that the tides typically ebbed and flooded only once a day instead of twice, which amazed him, since "I must needs confess it different from all that ever I observ'd in any other Port." Nor was there any consistent relationship between the highest tides and the full or new moon. And between these times of syzygy there were periods of stagnation or intermission, when the tides were either very small or non-existent.

Davenport wrote a letter to the secretary of the Royal Society in London detailing these anomalous tides, which had a maximum vertical range of about two metres (seven feet) but were especially erratic during the monsoon season. He warned navigators not to try crossing the bar at times of stagnation, because although the tide then was small, it was also highly unpredictable. To avoid running aground on the bar it was best to wait a few days for a strong high tide to carry the ship safely over the bar, and to seek help from a local pilot. He presented full information on the times, dates, and tidal variations as he had recorded them, and even offered a tentative explanation for the surprising phenomenon.

In London, the information came as something of a shock. As David Cartwright has noted, "Davenport's letter was disturbing to natural philosophers [what we today call scientists, a term not yet coined] seeking a rational mechanism for the tides in general, because the tides at Batsha broke all the rules known from observations on the coasts of Europe and further afield." Davenport suggested that the particular configuration of the Gulf of Tonkin, with its sandbar and several major inlets indenting the coast, might be in part responsible for the unusual tides.

The Davenport letter was passed along to Edmund Halley, a new member of the Royal Society and the astronomer who would later gain fame for accurately predicting the return of the comet that bears his name. Halley also expressed surprise at the letter, calling it "wonderful and surprising, in that it seemes different in all its circumstances from the *general rule*, whereby the motions of the Sea is regulated, in all other parts of the world I have yet heard of." Halley was busy with other projects; he may also have waited for confirmation of Davenport's observations, which eventually came from an East India Company captain who visited the Gulf of Tonkin in 1682. Finally, in 1684 Halley wrote a paper for the Society attempting a complex and jumbled explanation for the pattern shown by the Tonkin tides, the goal being to "assign a reason, why the *Moon* should in so particular a manner influence the *waters* in this one place."

This was several years before Newton published his *Principia*. Halley noted how little agreement there was about how the tides behaved, including those right along the shores of Britain itself, "of which we have had so long experience." For example, John Flamsteed, the Royal Astronomer, had published data on the tides as measured on the Thames at London and had written that "there is every where about England, the same difference betwixt the spring [high] and neap [low] tides that is here observed in the river Thames." But Halley contradicted Flamsteed, noting that tides on the river could not simply be applied to those at ports on the open sea. And, with possible relevance to the situation at Tonkin, Halley argued more generally that the way tides work in shallow seas and estuaries is different and more complicated than those on the open ocean. Later in his career, Halley would lead a scientific cruise that created the first detailed chart of tides and currents in the English Channel. He found, among other things, that tidal currents near the French shore ran at over five knots. No similar tidal chart of a major body of water was to be published for another 150 years.

Halley also played an intimate role in Newton's own efforts to understand and explain the tides. He became a confidant of Newton's and a leading popularizer and promoter of his theories. The *Principia* was published in 1687 at Halley's expense (he was independently wealthy). But it consisted of three volumes in Latin and was full of dense mathematics. It was not translated and published in English until 1729. Meanwhile, though, Halley summarized Newton's theories for King James II and wrote a detailed synopsis of the *Principia* in English for the Royal Society and the public at large.

While Newton was still working on the *Principia*, Halley apparently alerted him to the diurnal tides in the Gulf of Tonkin. Newton, realizing that this information ran counter to the main thrusts both of his theory and of observations in European waters, added a long paragraph to the *Principia* suggesting an explanation that was ingenious and suggestive of concepts yet to come, even if it has not quite stood the test of time. He proposed that the unusual ebb and flow at Batsha was the result of two separate tides flowing toward that port from different directions, one from the Indian Ocean to the south, the other from the South China Sea to the north. These tides, he thought, might be out of phase with each other. That is, the tide propagating from the Indian Ocean might take twelve hours to work its way through the channels leading into the Gulf of Tonkin, while the one from the South China Sea might take only six hours. This meant that the two tides would sometimes counteract each other, resulting in only one high and one low tide each day. Although he did not use the word *wave* for these tides, by implication Newton had hit upon the idea of wave interference, a concept that would play an important part in later theorizing not only about tides but also about light waves, sound waves, and the like.

William Dampier's Observations from Far-Flung Coasts

During these same decades, another Englishman — one whose contribution to the understanding of tides has been far too little recognized — was busy making careful observations of natural phenomena in many parts of the world. William Dampier has been aptly characterized in a recent biography as a "pirate of exquisite mind." Born in 1652 in Somerset, he went to sea at the age of sixteen, ended up in the Caribbean, and eventually joined the crews of various buccaneering ships that preyed on Spanish shipping and outposts, first in Central and South America and later in the East Indies. He became the first person to circumnavigate the globe twice, and ultimately did so a third time, commanding his own ship on the second voyage. Unlike many buccaneers who cared only for plunder, on all of his voyages he took detailed notes about the local plants, animals, and other natural phenomena. During his multi-year journeys, he successfully preserved those notes against the ravages of the sea, often by stuffing the precious papers into hollow cylinders of bamboo and sealing the ends watertight.

When Dampier returned to England after his first voyage, he began to share his observations with members of the Royal Society and spoke at least once at a Society meeting. In 1697 he published an account of his swashbuckling exploits in the form of a lively book, *A New Voyage Round the World*, which sold quite well. Encouraged by this success and the support of the Society, he published a second book in 1699, *Voyages and Descriptions in Three Parts*, which included a richly detailed third section titled "A Discourse of Trade-Winds, Breezes, Storms, Seasons of the Year, Tides and Currents of the Torrid Zone Throughout the World." This summarized his findings and provided Europe with its first overview of the complexity of tides across the vast oceans. It also showed how greatly they could differ from tides in the North Atlantic.

Dampier was careful to distinguish between tides (including tidal flows) and currents. "By *Tides*," he wrote, "I mean Flowings and Ebbings of the Sea, on or off from any Coast. Which property of the Sea seems to be Universal; though not regularly alike on all Coasts, neither as to Time nor to the height of the water. By *Currents* I mean another Motion of the Sea, which is different from Tides in several Respects; both as to its Duration, and also as to its Course." He went on to note that tides "alternately ebb and flow twice": that is, they "run forward, and back again, twice every 24 Hours: on the contrary, Currents run a Day, a Week, nay, sometimes more, one way; and then it may be run another way." Having observed the monsoon season in the western Pacific and Indian Oceans, he argued that currents were often linked to the winds, and added, "In some particular Places they run six Months one way, and six Months another," but in others "they run constantly one way, and never shift at all."

Another key difference he found was that tides are strongest close to the coastlines of very large islands and continents. "The force of Tides [and here Dampier is referring also to tide-induced currents] is generally felt near the shore; whereas Currents are at a remote distance; neither are the Effects of them sensibly discerned by the rising or falling away of the water, as those of the Tides are; for these commonly set along shore." As regards the vertical range of tides, Dampier was the first to point out a very significant pattern that is generally valid around the world: "I have also observed, that Islands lying far off at Sea, have seldom such high Tides as those that are near the Main." He cited several examples from his own experiences, such as tides at the Galapagos Islands, which he noted rise and fall a maximum of only about two-thirds of a metre (two feet), and at Guam, where the tidal range is not more than one metre (about three feet). By contrast, on the northwest coast of Australia he had found tides that rose and fell nine metres (thirty feet) or more, and in the Gulf of Panama, on the Pacific side of the Isthmus of Panama, about six metres (twenty feet). These macrotides were comparable to what was familiar to Dampier and his readers from the coasts of Britain and France.

Dampier offered no explanation for the difference between the tidal ranges at oceanic islands (such as the Galapagos, which rise up from a deep sea floor) and those on continental shores. So little was known about the depth of the oceans in Dampier's day that it probably would not have occurred to him that the size of the tide might be influenced by the way the oceans become much shallower as one approaches the continental shelves.

Another pattern that Dampier was apparently the first to notice was that the shape or configuration of a coastline could affect the tidal range considerably. In particular, where there were shallow and constricted bodies of water along a coast, such as lagoons or narrow river estuaries or inlets, he often found there to be relatively large tides. "This I have observed, that the greatest Indraughts of River[s] or Lagunes, have commonly the strongest Tides. Contrarily such Coasts as are least supplied with Rivers or Lakes [meaning in this case salt water or brackish lagoons partially cut off from the open sea by sandbars or islands] have the weakest Tides; at least they are not so perceptible," he wrote. "Where there are great Indraughts either of Rivers or Lagunes, and those Rivers or Lagunes are wide, though the Tide runs very strong into the Mouths of such River[s] or Lagunes, yet it does not flow so high, as in such Places where the Rivers or Lakes are bounded in a narrow Room, though the Tides do run of an equal strength at the Mouth or Entrances of either."

One of the examples Dampier cited for this generalization was several rivers that flow into a lagoon on the Gulf of St. Michael, which is part of the Gulf of Panama. "The Rivers that run into this Lagune are pretty narrow and bounded on each side with steep Banks, as high as the Floods use to rise.… The Lagune at the Mouth of the Rivers is but small, neither is there any other way for the Water to force it self into, beside the Lagune and the Rivers; and therefore the Tides do rise and fall here 18 or 20 Foot [5.5 to 6 metres]." Dampier's measurements were quite accurate. Modern measurements of the tidal range at the Pacific entrance to the Panama Canal show a rise and fall of 6.5 metres (21.5 feet).

Dampier contrasted this with the situation farther south, at today's Guayaquil, Ecuador. "The River of Guiaquil, in this respect, is much the same with the Gulph of St. Michael; but the Lagunes near it are larger. Here the Tide rises and falls 16 Foot [just under five metres] perpendicular. I don't know of any other such Places in all the South Seas; yet there are other large Rivers on the Coast, between these Places; but none so remarkable for high Tides."

The extraordinary British buccaneer also debunked a curious myth about tidal geography. This was the belief that the tides flow under the Isthmus of Panama between the Pacific and the Caribbean through "a Subterreanean Communication," making the isthmus "like an Arched Bridge, under which the Tides make their constant Courses, as duly as they do under London Bridge." Support for this idea allegedly came from the existence of "continual and strange Noises made by those Subterranean Fluxes and Refluxes; and that they are heard by the Inhabitants of the Isthmus." But Dampier would have none of it. "For I pass'd the Isthmus twice, and was 23 days in the last Trip that I made [bushwacking tediously] over it; but yet did I never hear of any Noises under Ground there."

Dampier was among the first to notice that the timing of tides elsewhere could be quite different from that on the coast of Britain. "We are taught, as the first Rudiments of Navigation, to shift our Tides, i.e. to know the time of full Sea [high tide] in any Place." This is essentially the same as knowing the establishment of a port, i.e. the relationship between the time of high water and the time the Moon passes the meridian. Dampier wrote that this is "indeed very necessary to be known by all English Sailers, because the Tides are more regular in our Channel, than in other parts of the World." Elsewhere, sailors have to adjust and observe the local tides, "which property of the Sea seems to be Universal; though not regularly alike on all Coasts, neither as to Time nor the height of the Water."

Part of Dampier's motivation in paying such close attention to the tides related to a highly practical matter. Sailing ships on long voyages had to be careened periodically to scrape and clean away the accumulated seaweed, barnacles, and other growth. Often there was also a need for repairs to the hull below the water line. Careening could be done where tides were small, but then part of the hull remained wet and difficult to reach throughout the cycle of ebb and flow. It was much easier and more efficient where the tidal range was large and the ebbing tide left the entire ship high and dry, at least for part of the twelve-hour cycle. "In all my Crusings among the Privateers," Dampier wrote, "I took notice of the risings of the Tides; because by knowing it, I always knew where we might best hall ashore and clean our ships: which is also greatly observed by all Privateers."

Where the tidal range was large enough and the timing was right, the situation was even better. A ship could be put onto the beach on a large spring tide. Then, as the tidal range decreased day by day, not even the high tide would wet the ship's bottom. The crew would have all the days of the neap tide to clean and repair the ship in dry, convenient conditions. In 1688, the *Cygnet*, on which Dampier was one of the more experienced of the crew, found the ideal circumstances during an extended stay at Kings Sound on the northwest coast of Australia. Dampier would not have been entirely surprised at the tidal regime there. The Kings Sound area happens to be almost the only part of the Australian coast where the tides behave quite similarly to those on the British side of the North Atlantic, being very large, mainly semi-diurnal, and quite regular. Modern measurements have found the Kings Sound to have the highest tides in Australia, up to eleven metres (thirty-six feet). As Dampier wrote, "We hal'd our Ship into a small sandy Cove, at a Spring-tide, as far as she would float; and at low Water she was left dry, and the Sand dry without us near half a Mile [almost a kilometre]; for the Sea riseth and falleth here about five fathom [thirty feet, or nine metres].... All the Neep-tides we lay wholly a-ground, for the Sea did not come near us by about a hundred Yards [one hundred metres]. We had therefore time enough to clean our Ship's bottom, which we did very well."

While on the Australian coast, Dampier had two full months to observe the local tides, and he noticed that the way their timing related to full or new moon was significantly different from in European waters. "In all the Springs that we lay here," he wrote, "the highest [tides] were 3 Days after the Full or Change, and that without any perceptible Cause in the Winds or Weather." This was contrary to what was thought to be the pattern in Britain, where, as we have seen, the highest tides in each cycle came only three tides, or about a day and a half, after syzygy.

This difference was highlighted when, as mentioned, the *Cygnet's* crew careened their ship to clean its bottom. Some, Dampier wrote, "were startled" at how the pattern of tidal timing differed from the European norm. Dampier himself had been observing the timing, but others had not. And so, "in that Spring [tide] that we designed to hall off, in order to be gone from thence, we did all take more particular notice of it than in the preceding Springs." Others in the crew had been paying less attention to the tides. "And therefore the Major part of the Company, supposing that it was a mistake in us who made those former Observations, expected to hall off the Ship the third Tide after the Change [new moon]." Three tides after syzygy would have fit the pattern that Colepresse and others had reported for British waters. "But our Ship did not float then, nor the next Tide neither."

This appeared to some of the crew like a serious problem. It "put them all into an amazment, and a great Consternation too: For many thought we should never have got her off at all, but by digging away the Sand; and so clearing a passage for her into the Sea." Trying to lighten the ship and dig a channel to the sea would have been an arduous task. "But the sixth Tide [i.e. about three days after syzygy] cleared all those doubts; for the Tide then rose so high as to float her quite up; when being all of us ready to work, we hall'd her off; and yet the next Tide was higher than that, by which we were now all thoroughly satisfied, that the Tides here do not keep the same time as they do in England."

Dampier reported on the range of tides, and their directions of ebb and flow, from dozens of locations around the world. But other than noting the tendency for tides along continental shores and in narrow estuaries to be larger than at mid-ocean islands or on open shores, he found no really consistent pattern in the distribution of tidal regimes. Tides in the tropics could be very large, as in northwestern Australia and on the Gulf of Panama. Or they could be very small, even along a continental shore, as on the Caribbean side of Panama, where they rise and fall barely one-third of a metre (about one foot).

French Tidal Science

The British were not the only nation to take a scientific interest in the tides. In 1701, two years after the publication of Dampier's treatise on tides, the Académie Royale des Sciences in Paris launched an effort to measure the tides that was similar to the Royal Society's long-term project. In France, Newton's gravitational theory of the tides was not yet generally accepted. Descartes's ideas about vortices in the ether were still considered persuasive as the theoretical explanation of how tides were generated. But the French acknowledged that there was still far too little precision about the exact pattern of tides along the coasts of France.

As a memorandum from the Académie noted: "Although the flood and ebb had been regarded as an impenetrable wonder, perhaps their cause has now been discovered, to the honour of Monsieur Descartes. But surprisingly, we have the System [of explanation] without assurance that the Phenomenon itself is known with sufficient exactitude." Therefore, the Académie "decided to obtain direct observations of the tides by able persons from various locations," in order to "profit from" them.

Some fairly accurate measurements of the times and heights of tides at the port of Brest in Brittany had already been made in the late 1670s. Now the Académie issued clear instructions on how additional data should be gathered. For example, tides were to be recorded over long periods of times. The people in charge of the effort were professors of hydrography, trained specialists assigned to each port who understood the sea and were capable of accurate timekeeping. The devices used to measure the height of the tides were graduated in feet and inches. At least one of them involved a float that rose and fell within a vertical pipe attached to the side of a wharf, with a long, lightweight rod attached to the top of the float. This made it easy to see where the tip of the rod stood against a graduated vertical scale.

The resulting measurements, not only from Brest but also from such ports as Dunkerque and Le Havre, were sent back to the Académie for analysis by Jacques Cassini, the astronomer who headed the Observatoire Royale. Because at many places observations were made both day and night, the long series of measurements were both accurate and useful in discerning subtle patterns in the tides. Between 1710 and 1720, Cassini noted, for example, that spring tides followed the full or new moon by about two tides near Brest, but by three or four at Dunkerque or Le Havre. He supposed that the difference was due to the tide moving eastward up the Channel, a concept that (like Newton's explanation for once-daily tides at Tonkin) anticipated the concept of tides propagating along a shore or channel as progressive waves.

The long series of measurements allowed Cassini to tease out some of the less obvious details of tidal phenomena. By the early eighteenth century, British measurements at Plymouth and Bristol had discerned a slight diurnal inequality. (At certain parts of the fortnightly cycle, one high tide might differ from the preceding or succeeding one by about one-third of a metre, or one foot.) The higher of each day's two high tides, for example, usually came in the afternoon in summer and in the morning in winter. Cassini found a similar pattern on the French coast. (At first, this was thought to be a general rule, although future data from other ports revealed a more complex pattern.) And he found, as the British had done, that tidal ranges were larger when the Moon was at perigee. In addition, he noticed that if he allowed for the Moon's varying distance from the Earth, he could discern a separate pattern in the size of tides that varied with the Earth's distance from the Sun, or what astronomers call its parallax. During Cassini's day, the Earth was closest to the Sun, a position called perihelion, in late December, close to the time of the winter solstice. (The Earth's perihelion shifts, or precesses, slowly. Today perihelion comes more than a week later, in early January.) What Cassini noticed was that spring tides were also larger then than they were at the summer solstice. This was the first conclusive demonstration of parallax.

The mid-eighteenth century was the high point of the Enlightenment in Europe. Unbiased observation and rational thought had begun to win out over religion, tradition, and revealed knowledge. So it should not be surprising that by that time the study of the tides was gradually becoming an orderly and systematic science. Progress in understanding the tides came only very slowly, however. Each discovery in the realm of empirical measurement seemed to pose new questions and begged for a theoretical explanation. On both sides of the English Channel, tidal theory lagged behind the practical knowledge of how the sea ebbed and flowed. It would be decades before new ideas and concepts brought theory into much closer line with practice.

Tidal Tempest

7

Fast-Flowing Waters Doom Many a Ship

Skookumchuck means "place of strong waters" in the Chinook coastal jargon, the traditional lingua franca of the Natives on the northwest coast. And the spot named Skookumchuck Rapids, located along the British Columbia mainland shore north of Vancouver, fully earns that designation. Hiking in from the road with a friend one time, I first noticed a distant rumble pervading the still forest air. Then the wooded trail opened up and we came to a cliff overlooking a raging channel of white water. Off the nearest point of land, jagged standing waves rose some two metres (about six or seven feet) into the air like sharp teeth. In other places, water plunged into deep, gurgling holes. Swirling green bands of turbulence were swept along with the flow and spun off into powerful back eddies where they were deflected by the rocky shoreline.

In mid-channel, the torrent gushed through a gap between small islands and fanned out into a plume of milky rips as it rejoined the main stream. Seagulls wheeled and swooped over the tumult and foam, hunting for fish that had been injured in the natural cataclysm or simply

Surfing the standing waves of the Skookumchuck Rapids in British Columbia.

thrust to the surface by the powerful upwellings. It was like an angry river, but the water was salt, not fresh. And rather than plunging down a steep mountainside, it was being propelled by the ebb and flow of the Pacific tides.

Skookumchuck Rapids has some of the fastest tidal currents in the world. It is one of a score of places on the Inside Passage where swift tidal flows put on spectacular displays, create special conditions for wildlife, and give mariners sleepless nights. It is the shallow, constricted entrance to a network of fjords (locally called inlets) that cut deep into the coastal mountains on the eastern side of British Columbia's largest body of sheltered water, the Strait of Georgia, which separates Vancouver Island from the B.C. mainland. Also called Sechelt Rapids, Skookumchuck is like the narrow neck of a very large bottle that is repeatedly filled and emptied as the tides rise and fall.

As the tide floods into the nearby Strait of Georgia, hundreds of billions of gallons pour through the narrow channel into Sechelt, Narrows, and Salmon Inlets, and the show begins. After about six hours, the water level inside the nearly eighty kilometres (fifty miles) of inlets has risen as much as three metres (ten feet). The flow slows to a trickle and goes slack. Then, after only ten or fifteen minutes of relative calm, the tide out in the strait ebbs, and all the water in those inlets begins to rush back out again. At peak spring tides, it surges through some places in Skookumchuck at more than sixteen knots.

Skookumchuck is only one of many tidal rapids on the B.C. coast. Farther north along the mainland coast is Nakwakto Rapids, which is the gateway between the open waters of Queen Charlotte Sound and nearly landlocked Seymour and Belize Inlets. There, too, the water can race through at around sixteen knots, and a much larger volume of water moves at this high speed than at Skookumchuck, so the display is in some ways even more spectacular. In the middle of Nakwakto is Turret Rock, a tiny islet also known as Tremble Island. A government surveyor once decided to sit out a tidal cycle there. The islet shook so violently from the force of the current that he lay face down, clung to the tufts of grass, and stuffed his ears against the roar. Seen from the air when a big tide is running, Tremble Island looks like a ship plowing into turbulent seas. It kicks up a huge white bow wave on the end that is being hit by the tidal flow and leaves a long, swirling wake of turbulence in the other direction. Even the orcas (killer whales) that frequent the region negotiate Nakwakto only near slack tide.

There are also very swift and powerful tidal currents (up to eight or nine knots) at several places in the Puget Sound area of Washington State, and along the coast of Alaska. Some of the names in these areas are highly evocative of the local conditions, such as Deception Pass and Peril Strait. By way of comparison, the currents that sweep in and out of San Francisco Bay at the Golden Gate are infamous for their swiftness and the hazardous sea conditions they can cause. When people jump from the bridge to commit suicide, their bodies are frequently swept away and never retrieved. The maximum speed of those currents, though, is "only" around five knots.

Tidal rapids occur not only at the mouths of mainland inlets but also in the numerous narrow passes among the myriad islands between Vancouver Island and the mainland. The tide from the open Pacific comes around both the northern and southern ends of the mountainous, five-hundred-kilometre (three-hundred-mile) long island and into the inner waters of Queen Charlotte Strait, Johnstone Strait, and the Strait of Georgia. At some places, it does not arrive at both ends of a channel at the same time. At Seymour Narrows — between the islands of Vancouver and Quadra — the discrepancy can be two hours. As a result, the difference in water level (also called the hydraulic head) between one end of the channel and the other often exceeds one metre (three feet) and generates currents that can run at more than fifteen knots. The enormous volume of water flowing through this gap, and the fact that the channel is part of the main shipping route to Alaska, makes Seymour Narrows the most dangerous tidal rapids on the coast. That's the channel where I was almost swamped by the bow wave of a large fishboat while out in a small runabout with a fishing guide. But there are rapids with top current speeds of eight knots or more on many inter-island channels, and all are places that mariners have learned to treat with the utmost respect.

The Strait of Magellan

As we have already seen, swift tidal currents can create powerful whirlpools, such as the infamous Maelstrom off Norway, the Naruto Whirlpools of Japan, and the Corryvreckan in Scotland's Hebrides. Many places in the world have considerably larger tidal ranges and swifter currents than those places, including the coast of northwestern Australia and the south and west coasts of Korea. Huge whirlpools are not always the result. Still, the currents themselves and other effects caused by them can wreak plenty of havoc.

Ferdinand Magellan's fleet, for example, set out on history's first circumnavigation in 1519. They soon found themselves buffeted by the extreme tides and wild tidal currents of farthest South America. In the long and tortuous Strait of Magellan, tidal currents can run at twelve knots or more. Magellan made it through the maze of narrow channels with a relative speed and ease that were not often repeated. Within the strait, fierce and unpredictable catabatic winds or williwaws, caused by cold air cascading down off the towering mountains and glaciers, were one of the hazards. But the main problem was the powerful tidal currents, which were so treacherous that the Spanish came to regard the strait as a secure natural line of defence against British attack on their growing Pacific empire.

Sir Francis Drake proved them wrong by making his way through the strait in 1578. He did this largely by exploiting the tidal currents to hasten him along when their direction of flow was favourable and anchoring to wait out the tide when it was not. Thus he managed to surprise the Spanish, plundering towns and shipping on the Pacific shores of South and Central America and proceeding at least as far north as California. But he also discovered the alternate route between the Atlantic and Pacific farther south, around the outside of Cape Horn. So many later European ships came to grief in the Strait of Magellan that most future shipping preferred to risk the dangers of rounding the Horn.

The tidal range in the eastern end of the strait, and just outside it on the Atlantic shore of Patagonia, averages about nine metres (thirty feet) and sometimes reaches about fourteen and a half metres (almost forty-five feet). On the western end of the strait the tidal range is around six metres (twenty feet). Tides in the southern Pacific circulate and butt up against the coast of Chile according to one tidal regime, while those in the southern Atlantic follow an entirely different one. Separating the two is the huge island of Tierra del Fuego and all its outlying islands and shoals, right down to Cape Horn. The timing of the tides propagating into the strait from one ocean is quite different from that of the tides coming from the other side. As with the tides at Seymour Narrows in British Columbia, the difference in water level at the two ends of the Strait of Magellan can be enormous. Whenever there is such a large hydraulic head, water will flow rapidly "downhill," seeking to equalize the levels, which is what generates the fast currents.

Sailing ships were often stopped dead by those currents, even when the winds were favourable. As Drake did, they had to claw their way through the strait, day by day, anchoring whenever the currents were contrary if they could find a suitably sheltered and shallow spot. It often took them months to make it through, and many were simply swept onto rocky shores by the currents and wrecked. But even many modern steamships, with all the advantages of predictable and steady engine power, have come to grief there.

The Nova Scotia–born American sea captain Joshua Slocum was the first person to circumnavigate the world single-handedly, and he wrote a colourful and exciting book about his experiences. When he sailed his sloop *Spray* through the strait in 1896, he had to use tactics similar to those of Drake. Without an engine, and with no crew to relieve him at the wheel when he got weary, Slocum fought an endless battle with the tides. From the moment he entered the strait, the tides and their effects were among his greatest concerns. "As the sloop neared the entrance to the strait I observed that two great tide-races made ahead, one very close to the point of the land and one farther offshore. Between the two, in a sort of channel, through combers, went the *Spray* with close-reefed sails. But a rolling sea followed her a long way in, and a fierce current swept around the cape against her."

Slocum had one advantage over large ships, however. His boat was small enough that if he could not find a good place to anchor, he could simply moor his boat to thick kelp beds whenever the currents ran against him and wait out the tide. Then he could ride with the tidal currents when they again became favourable. Slocum went ashore frequently and found grim testimony to the navigational hazards. Even in the late nineteenth century with its steam power, so many ships had been destroyed in the strait that Slocum was able to salvage great quantities of still-usable wreckage, including casks of tallow and a barrel of wine. As he noted at one point, "the wreck of a great steamship smashed on the beach abreast gave a gloomy aspect to the scene." He also came upon the graves of dozens of sailors who had perished in that loneliest part of South America.

Even today, tidal currents make the area extremely dangerous for navigation. In October 2003 the American guided missile frigate *Robert G. Bradley* became the first U.S. Navy ship to pass through the strait in eight years, starting from the Chilean end. The commanding officer, Michael Strano, called the passage "an experience the crew will never forget." The navigation was so tricky that another officer, Richard Koch, commented, "I've never done anything this intense, and I've been in the Navy twenty-one years. The currents are so bad in some areas, they can run you into the cliffs." As the shipboard newspaper reported, "The transit speed had to be adjusted to arrive at the most narrow part, the English Narrows, which is only 90 yards [82 metres] wide, at slack water with no current. With mountains dwarfing [the *Bradley*] on either side, the water was almost 600 feet [183 metres] deep. That day and night [the *Bradley*] continued steaming southward, often reducing its speed to as little as 12 knots for maneuvering." It continued, "As the ship passed through the two 90 degree turns in the English Narrows, on the starboard side was a statue of the Virgin Mary. Tradition calls for sailors to toss a coin at the statue as the ship passes and make a wish for safe passage. Many coins were tossed that day."

In the late 1990s, a tugboat engineer, Keith Lewis of Block Island, New York, experienced somewhat different hazards in the area when he did a stint towing oil and gas drilling rigs in and around the Strait of Magellan, now a centre for petrochemical exploration. But all those hazards were due to the tides or resulting currents. "A lot of water moves in and out each day," he recalled, "driving swift currents that have defeated many a ship. Those were the waters on which we worked." When the rig was in place and drilling, the tug was expected to hover close by when the roughnecks were in danger of falling overboard, especially at night. In fact, two men did fall, and they perished. "With the darkness, cold water, and swift currents, there was little chance of retrieving them alive."

The two ports in the area, said Lewis, were primitive and difficult to access. "Those thirty-foot [nine-metre] tides governed our use of the ports. The entrance to both had sandbars over which we had to pass; we could cross only when the tide was relatively high." At one of the ports a large pier had been built to accommodate a government tanker, but "it was of little help for vessels our size [which at low tide would have tied up far below the level of the pier]. With the extreme tides, we could not get on and off the boat except at high water."

And even once a vessel had made it over the bar and into the harbour, the waters remained treacherous. One time, a Norwegian supply boat came in and tied up astern of Lewis's tug. Powerful currents swirled around the legs of the pier, creating whirlpools and eddies. There was a heavy strain on the mooring lines as they made fast. Their chief mate was handling one of the lines; it momentarily caught on a fitting, then jumped free. The line sprang up, hit the mate in the neck, and decapitated him. But the pace of work never faltered. Out at the rig "that drill bit continued round and round, cutting deeper … in the quest for black gold. Yes, time is money; death is lonely in the oil patch."

Pentland Firth

Another area where incredibly fast tidal currents have brought many ships to grief is Pentland Firth on the northern coast of Scotland. The firth is actually a strait just over ten kilometres (six miles) wide between the mainland of Scotland and the Orkney Islands. The tides flow through between the Atlantic and the North Sea generating currents that have been reported, at one spot, as running at up to sixteen knots, which would make them equal in speed to the fastest tidal rapids on the coast of British Columbia. In many other parts of Pentland Firth the currents have been measured at nine to ten knots.

As on the west coast of Canada, the currents in Pentland Firth produce treacherous rips, overfalls, and whirlpools. Some of these localized sea conditions are so notorious that they have acquired colourful names. One, called the Merry Men of Mey, has been described as a natural breakwater of high standing waves, a line of heavy breaking seas that extends right across the firth. Another, the Swilkie, is particularly violent and often generates an awesome whirlpool by the same name, which derives from an Old Norse term meaning "the swallower." The local advice to mariners states that the Swilkie is "dangerous and should be avoided even in fine weather." It is most violent when a westward-flowing tidal stream is opposed by a gale blowing from the west.

Adding to the dangers in the Pentland Firth are two islands, Stroma and Swona, that straddle the main channel through the strait, as well as a nasty series of rocks, called skerries, that lurk smack in the middle of the strait's eastern end. The area is also prone to dense fogs. Especially before the days of radar and precise satellite navigation, countless lives were lost as the tides swept ships to their doom on those hostile shores and rocks. In fact, so many ships have come to grief over the centuries that the inhabitants of the islands became notorious as wreckers. Besides fishing and keeping sheep, they derived much of their living by illegally salvaging the booty that wound up on shore, and were often accused of ignoring the plight of shipwrecked people as they raced to grab and hide the valuable goods.

On the northern side of Pentland Firth is the southern entrance to Scapa Flow, a huge but sheltered wartime fleet anchorage that is tucked in among the Orkney Islands. It served as the main outpost for the Royal

Navy during both world wars, largely because its location enabled the British to prevent ships coming from Germany or Scandinavia from gaining access to the Atlantic. There were many natural channels leading into the anchorage. To protect the navy's vessels inside, some of these channels were closed off to shipping by minefields, others by old sunken blockships and barges that had been placed there intentionally as obstacles. Other channels had floating booms across them, which could be opened briefly to allow friendly ships to come and go, and submarine nets were strung between the islands. In addition, the vicious tides of Pentland Firth and the other approaches to Scapa Flow were thought to be a major deterrent to any intrusion by enemy ships, especially submarines. As one historical account put it, "any prospective attacker had to deal with not only the heavy defenses, but also the unpredictable currents, powerful enough to carry a U-boat off course and into danger." The Royal Navy thought its base was essentially impregnable.

Then, at the beginning of the Second World War, the commander of Germany's U-boat fleet, Admiral Karl Doenitz, realized that in one of the channels leading into the Flow there was a narrow gap between two of the blockships and just enough water depth for a U-boat to get in by running on the surface. He also calculated that if he picked a new moon night, there would be darkness and an extremely high spring tide. The U-boat would have shallow enough draught that it might be able to slip in between the blockships and over the cables strung between them. If the timing were just right, slack tide would mean greatly reduced currents as well.

Doenitz selected his most audacious U-boat skipper, Gunther Prien, for the raid. October 13, 1939, brought one of the highest tides of the year. U-47 manoeuvred though the supposedly blocked passage, scraping along over the steel cables and briefly running aground on a sandbar. Once he got free, Prien found the battleship *Royal Oak* riding at anchor, unaware of any danger. He sank the huge ship with torpedoes that hit its aft magazine, causing a giant explosion and sending more than eight hundred sailors to their deaths. As one of the opening salvos in the naval war, it severely shook British confidence.

The largest tides and most powerful currents in American waters are on the coast of Alaska, where Cook Inlet has a maximum tidal range of more than twelve metres (about forty feet). The inlet is a long fjord with a river flowing into it. Initially thought to be the estuary of a river, it was at first called Cook's River. Captain Cook discovered it in 1778 on his third voyage, when he was searching for a possible western entrance to the ever more elusive and unlikely Northwest Passage between the Atlantic and Pacific.

Cook's ships proceeded up the inlet for several days, riding the tidal currents against unfavourable winds. As described in the journal kept by surgeon's mate David Samwell, Captain Cook made progress by "coming to an anchor every ebb tide & weighing with the flood." Eventually the inlet divided into two arms. Cook sent out longboats to explore further, and they reported back that the inlet dead-ended among tall peaks. There was no Northwest Passage opening onto this part of the Alaska coast. It was time to turn around and exploit the tidal currents in the opposite direction. The next day, Cook "weighed with the first of the Ebb and with a gentle breeze at South plyed down the River, in the doing of which, by the inattention and neglect of the man at the lead, the Ship struck fast on a bank that lies nearly in the Middle of the River." But unlike Cook's mishap years earlier, when he'd run aground on a reef off Australia and taken on a dangerous leak, in Alaska he came away unscathed. "At the flood tide made [sic] the Ship floated off without receiving the least damage." The next few days they continued to ride the favourable tides, "working with the Tides down the River having contrary Winds all that time," until they reached the open Pacific.

Farther south along the Alaska coast is a glacial inlet called Lituya Bay that is particularly hazardous because of the tides. In 1786 the French explorer Count La Pérouse piloted his two square-rigged ships warily into the narrow entrance to the bay, but found himself being thrust forward in the grip of a powerful current that generated nasty rips and nearly drove him onto the rocks. "In the 30 years I have been sailing," he wrote, "I have never seen two ships so close to destruction." Modern measurements show that the current runs through that narrows, today called the Chopper, at up to twelve knots.

During his fleet's stay in Lituya Bay, La Pérouse discussed the tricky conditions with the local Tlingit people. "We learned that seven very large canoes had lately been lost in this passage, while an eighth escaped.... These Indians seemed to have considerable dread of the passage, and never ventured to approach it, unless at the slack water of flood or ebb."

Not even this useful intelligence averted tragedy, however. After spending more than a week charting the sheltered bay, which he named Port des Francais, La Pérouse sent three of his boats to take soundings near the entrance, warning them to stay away from the channel if they saw breaking waves. But the boats got too close and were drawn by the ebbing tide toward the narrows at what a survivor described as "extreme velocity." Rowing hard, the crew of one boat escaped. The others were sucked into the churning nightmare, where both boats were swamped and swept out into the Pacific. All twenty-one crewmen died. The French and Tlingit searched the rocky shores outside the bay, but never found any remains. La Pérouse had journeyed halfway around the world "without having had a single person sick" until this disaster struck. "Nothing remained for us," he lamented, "but to quit with speed a country that had proved so fatal." His luck had apparently run out. After visiting Australia, his ships were wrecked on reefs in the Santa Cruz Islands, and none of the survivors ever made it back to Europe.

Even in more modern times, boats and ships enjoying the advantages of powerful engines have frequently come to grief in the tidal rapids of the northwest coast.

In 1923, the steam-powered tug *Peggy McNeill* was towing two coal scows and had to navigate through narrow Porlier Pass, which separates Galiano and Valdes Islands on the inner coast of British Columbia. Peak currents in Porlier Pass can run at nine to ten knots. The daughter of the local lighthouse keeper recalled after the event that "it was quite a calm night, although it was a strong, strong tide, and Daddy and I watched her come down the inside shore ... then we went to bed." W. Ingram, the mate on the tug, later reported, "The sea was really boiling when the tug entered the riptide of the pass.... We managed to get a short distance past the second light when the tug suddenly took a shear and the hawser [tow line] started to go over the rail." Ingram ran for an axe to cut the line before it could drag the tug sideways, but he was too late. The scows, in the current's grip, took control and rolled the tug right under. Ingram escaped from the upside-down tug before it sank and swam hard to catch a rope trailing from one of the scows. He was sucked under briefly by a whirlpool, climbed onto a bell buoy to rest, and eventually swam to a scow and climbed aboard. Some of the other men managed to grab onto a rope from the other scow, but Ingram was unable to help them. "The waves were plunging the scows up and down, making it difficult to hold on. These men had been in the water for almost three quarters of an hour and were losing their strength, for one by one they disappeared." Five men perished that night.

By far the worst of the tidal channels in terms of loss of life over more than a century is Seymour Narrows. Peak currents there run at fifteen or sixteen knots. Situated on the main navigational channel between Puget Sound and Alaska, it is the route taken by all large ships, and it used to be obstructed by a huge, sunken rock. Until Ripple Rock was destroyed by blasting in 1958 (in the largest peacetime non-nuclear explosion to that date), at low tide its twin peaks reached to within about three metres (ten feet) of the surface in the middle of the channel. The rock created nightmarish conditions when the currents swept around and over those submerged peaks.

Mike Frye, a fisherman, never forgot his first trip past the rock. "I saw this wild whirlpool from a distance. I could see some logs as they whirled around and around at a tremendous speed and I was tempted to get a closer look ... but I soon chickened out ... I could feel the boat being drawn into it." He got away, "but I got a look at it, and it was some sight ... I don't know how long that log was, but I saw twenty or more feet [six metres] sticking out of the water, straight up mind you and going like hell."

Curiously, when Captain George Vancouver first sailed through Seymour Narrows in 1792, he must have had a slack tide and hugged one side of the channel. He failed to notice Ripple Rock and reported the channel safe for navigation. Later ships were less fortunate.

The rock's first major victim was the U.S. Navy sidewheel steamer *Saranac* in 1875. Seaman Charles Sadilek described Seymour Narrows in his diary. "Here the tide is forced through a narrow, winding channel. There are foaming swirls over the face of the rocks and great eddies caused by meeting currents. A deviation of a point [just over eleven degrees] from the true channel sends a ship to destruction," he wrote.

"Here the contending currents take a vessel by the nose and swing her from port to starboard and from starboard to port as a terrier shakes a rat. It may be doubted if the Argonautic expedition experienced greater perils than are to be met in Seymour Narrows."

Sadilek placed the blame on the captain's unwillingness to wait for the tidal conditions to be right. "The charts warn every pilot and captain against attempting a passage through at low tide. They also point out a dangerous rock near the center, which becomes doubly so at such times. The pilot had on many former occasions guided steamers through in safety, but always at flood tide. If reports are true, he attempted to persuade the captain of the *Saranac* not to venture through at low tide but the captain answered, 'I'll risk it,'" he recorded. "The writer happened to be on deck as we neared the Narrows and heard the pilot order the engineer to put on all possible steam, intending undoubtedly to rush by the dangerous rock in safety by using great speed. When in the midst of the whirlpools the ship refused to answer her helm, and was for a moment beaten about by the angry waters when all of a sudden there came a crash and shook the ship as if it had been fired into by a battery of guns.... The mad currents had driven the *Saranac* on one of the rocks which crushed a hole in her side." The ship was wrecked, but miraculously, all the crew were saved.

Ripple Rock eventually sank or severely damaged at least 14 large ships and 100 smaller vessels, and took at least 114 lives. After the colossal blast destroyed the peaks of the rock, the highest point was a safe fourteen metres (forty-six feet) below the water at low tide, but the government official in charge warned, "Seymour Narrows is still narrow. There will still be heavy tides and currents, so navigation will still be highly hazardous."

This fact was underscored in June 1984 when the luxury cruise ship *Sundancer*, bound for Alaska with eight hundred passengers on board, ran onto the rocks along the side of Seymour Narrows and slashed open its bottom. A Coast Guard investigation attributed the accident to mistakes on the bridge as the 153-metre (502-foot) ship, running north in darkness at twenty-two knots, closed quickly on the Narrows in the grip of a ten-knot north-flowing current that tended to set the ship to starboard. The helmsman may have misunderstood one command. Then the pilot apparently misspoke himself and ordered a course change *toward* the starboard shore. The captain and pilot soon realized the error and took drastic evasive action, but it was too late and the *Sundancer* struck amidships. Leaking badly, she developed a list, and panic broke out among the passengers as they fled their cabins and crammed the stairwells to the lifeboat stations. Water flooded in quickly, but the crippled ship managed to limp across the Narrows and tie up next to a pulp mill wharf. Efficient rescue efforts prevented any loss of life, but the ship sank in the shallows and destroyed the massive wharf, for a total loss of over $60 million.

The awesome show put on by British Columbia's tidal currents, and the dangers they pose, eventually attracted people who wanted to challenge them just for sport. Today these are mainly kayakers, who like to surf through Skookumchuck Rapids on big tides. They wear helmets, and their nimble little kayaks, sealed up with tight spray skirts, are essentially unsinkable. When they hit the high standing waves they routinely go right over and are engulfed in foam, only to bob up and right themselves after a few seconds. It's a thrilling extreme sport, and one that dismays some of the local fishing guides, who worry that less skilled people, especially sport-fishing types, will also treat the rapids without due respect. And then it is the sport-fishing guides who will have to risk their lives by coming to the rescue when the inevitable accident occurs. "If there's one thing I'd like to emphasize," one guide told me, "it's that this is not kayaking water. It's *dangerous*."

Far better, then, to enjoy these natural spectacles from the safety of the shore. But even the wildest rapids have their placid moods. On another hike to Skookumchuck, I arrived to find it as smooth as the proverbial millpond. Not even a breeze stirred the water. It was dead slack. But only for a few minutes. Almost imperceptibly, the water began to flow. Streamers of bull kelp reached out and waved. The birds started to flutter. Skookumchuck hummed a lullaby, which soon became a sprightly march and eventually built to another roaring crescendo. And it was all being orchestrated by the rotation of the Earth and the attraction of the Moon and Sun in the firmament.

Theory Meets 8 Practice

Tides as Great Rotating Wave Systems

One summer in the early 1970s, I spent a few weeks taking a course at the University of Essex in Colchester, an ancient city that had been a Celtic stronghold in pre-Roman times. When there was free time, I enjoyed walking down to the nearby seaside in the small town of Wivenhoe. Its main attraction was a concentration of quaint old drinking establishments. Pub-crawling on a summer's evening was a popular student sport. Most of the pubs were located on the town's busy and scenic harbour. This was situated along the estuary of the River Colne, which empties into the North Sea not far from the port of Harwich, where the ferry runs across to Holland.

As I ambled along the waterfront, after quaffing one pint of good English bitter and before indulging in the next, I cast my sodden gaze out at the boats. Over the course of several such outings, I noticed a few things. If the tide was high, numerous small yachts bobbed at their moorings just off the town's shore, their colourful pennants flapping in the breeze. It was an enchanting tableau. When the tide was out, though, it seemed to be out a *very* long way. The tidal range in the area was sizeable, about three metres (ten feet) or more, and the estuary cut through the low-lying countryside of East Anglia. Low tide exposed broad mud flats, which, scattered with debris such as old rusty chains and encrusted concrete mooring blocks, made for a much less attractive scene. Out on those flats the moored yachts sat on the mud, and I could see that many of them were obviously designed to do so.

In my North American experience, most boats have a single keel along the centreline of the hull, and if it's a deep keel, as with a sailboat, that would cause the boat to lean far over sideways when aground. But these boats had twin sets of bilge keels attached well out from the centreline on each side. These allowed the boats to remain upright even at low tide. I had never seen bilge keels before, and I realized that they were a very useful feature for places like coastal Britain with large tides and extensive mud flats.

A year or so later, I was in Germany. Friends in Düsseldorf decided to take me on a day trip to the sea along the coast of Holland. When we got to the beach, I was struck by how similar it looked to beaches I had visited on the east coast of the United States, such as Jones Beach on Long Island, New York, or at Atlantic City, New Jersey. The waves were smaller, but the fine sand looked the same and the overall slope and configuration of the beach, with its grassy raised berm at the back, was almost identical. The tide was out, but, judging by where bits of seaweed and driftwood marked the high tide line, it was clear that, as on the middle Atlantic coast of the U.S., the tidal range was only about one and a half metres (five feet).

This surprised me, because I knew we were almost exactly due east of Wivenhoe, along the line of latitude 52, and only about 130 kilometres (80 miles) away across the North Sea. We were on the shores of the same body of water, and yet the tides on this part of the Dutch coast were only half as large as those on the coast of East Anglia. I wondered why.

William Whewell and the International Tidal Experiment

In the 1830s, British astronomer and mathematician William Whewell of Trinity College, Cambridge, also became aware of how different the tides were on the opposing shores of the North Sea. And it was not just a matter of the size of those tides. Whewell, who is best known for having coined the word *scientist* (in place of the term *natural philosopher*, which was used until then), noticed that tides in the North Sea behaved quite differently from the way he had expected, and he had been examining them very closely.

In 1829 it was discovered that the existing tide prediction tables for London were inaccurate. These had long been published privately for profit and were based on secret methods of observation and calculation. One of Whewell's colleagues at Trinity College, fellow mathematician and astronomer John William Lubbock, looked into the problem and attempted to improve the tables for coastal Britain, but he was having great difficulties. He had found that by observing the rise and fall of the tides over many years at a few key ports, useful local tide predictions for those specific places could be made. Lubbock thought that this approach, if carried out with great care and over a long enough period, was the key to unravelling the overall complexity of the tides. The trouble was that it did not, in fact, seem to be leading to an ability to predict the tides elsewhere. Nor did it bring Lubbock closer to a general theory that accounted for the great variation in the pattern of tides at different ports. Whewell decided to take up the challenge of "tidology" himself.

Whewell's approach to tidal science has been studied by science historian Michael Reidy of Montana State University, who notes that gathering numerous observations at a few places had proved to be inadequate. "Indeed, Lubbock had ... contented himself with studying the tides from observations made for many consecutive years at the London and Liverpool Docks." Whewell had a different concept. Instead of making or collecting long series of observations for only one or two places, to deduce the pattern of tides over time, Whewell felt he needed "comparative observations at different places," and in fact at *many* places. This would allow him to see how tides varied according to the relationship of those places in space. It was crucial, he thought, to see the geographical patterns as well as the purely temporal ones.

This was not entirely new. The Venerable Bede had long ago noticed that high tide progressed over time along major areas of the British coast. And the French astronomer Jacques Cassini had supposed that the tide propagated northeastward along the English Channel, with high tide coming earlier at Brest than at Le Havre or Dunkerque. But Whewell was seeking a much more complete and accurate set of data. As Reidy summarizes Whewell's project, "his ... approach ... entailed constructing a theory of the tides based on their progression along the entire coast of Great Britain. He could then extrapolate from one port to the next, and eventually to all ports in Europe and beyond. To attain such a theory of the progression of the oceanic tides, short-term observations at each and every port were needed," not just long series of observations over many years, as Lubbock had been collecting. And to connect the geographical data with the temporal, those observations had to be made simultaneously during a single, clearly defined time period. That was not something any single researcher could do.

Whewell received enthusiastic help from the Hydrographic Office of the British Admiralty, particularly from its head, Admiral Sir Francis Beaufort, the pioneer of weather forecasting, whose name is well known to mariners for the Beaufort scale of wind and wave conditions. Beaufort got the Admiralty as a whole to pitch in, including officials at naval dockyards in many ports. Whewell drafted detailed instructions on how observations were to be made, and Beaufort distributed those as orders to every Coast Guard station on the British Coast, some of them only a few kilometres apart. Each station was to record its observations of the tides and return the results directly to Beaufort.

By the mid-1830s, accurate tide gauges had greatly simplified the task of measuring and recording the height of the tide. Measuring time precisely was no longer a great problem, either. As Whewell himself wrote, "The Coast Guard people, it appears, are so regular in all their proceedings that time to them is essential and they are therefore tolerably well provided with watches, etc." And so, in June 1834, more than five hundred stations on the coasts of Great Britain and Ireland measured the height of the tide every fifteen minutes, and they did so day and night for a period of two weeks, a full fortnightly tidal cycle. This produced a vast database.

Whewell and Beaufort wanted to repeat the experiment a year later to check on the accuracy of the first set of observations, but then Beaufort proposed an even more ambitious idea. "Would it not be a delightful appendage to the batch of Coast Guard Tides which are to be observed this year," he wrote to Whewell in February 1835, "if we were to procure simultaneous observations along the shores of Holland and France, Newfoundland, Nova Scotia and [elsewhere in] North America?" Soon, with the support of the Foreign Office, the Admiralty was approaching foreign governments, and the responses from abroad were uniformly favourable. It was a period of international peace. Around Europe and across the North Atlantic, governments committed themselves to the first large-scale, multinational, cooperative scientific experiment in history.

In June 1835, for more than two weeks, some 650 tide stations took part and then returned the information to Britain. As during the year before, more than 500 of those locations were in England, Scotland, and Ireland. But there were also reports from 28 stations in the United States, 16 in France, 7 each in Spain and Portugal, 5 in Belgium, 18 in the Netherlands, and 24 each in Denmark and Norway.

Whewell then took on the daunting task of trying to discern some kind of order in this mass of data. By the nineteenth century, tides had come to be viewed as waves that progressed along a channel or across an ocean. Whewell set out to map the progression of those tide waves over time by drawing "co-tidal" lines. Each line connected all points where the tide was high at the same hour of the day (in Greenwich time), and there was a separate line for each of the twelve hours in a semi-diurnal tidal cycle. As he wrote in a letter to a friend, "I shall now have a register of the vagaries of the tide-wave for a fortnight as has never before been collected, and, I have no doubt, I shall get some curious results out of it." But he had no idea just how curious.

The assumption at the time was that, because the Earth rotated eastward, Newton's tidal bulges, moving westward, would set up a tide wave that progressed in a westerly direction. It would be blocked or deflected by the continents, except in the so-called Southern Ocean, where, it was thought, it could propagate without obstruction around the world. From that region, tide waves could also branch off and propagate northward into the world's major oceans, the Atlantic, Pacific, and Indian. The expectation was that in the North Atlantic high tide would reach both sides of the ocean at roughly the same time and then propagate into each of the marginal seas around its shores.

With reports from stations not only in Britain but also in Holland, Denmark, Norway, and even such Hanseatic city-states as Hamburg, the tidal experiment of 1835 gave Whewell a particularly good set of data for the shores of the North Sea. He expected to find that the tide wave, after coming northward up the Atlantic, would eventually penetrate to the North Sea. The English Channel was so narrow that it might block the tide. Instead, Whewell imagined that the tide wave would round the northern end of Britain and propagate southward through the North Sea, with high tide hitting both sides at approximately the same time. If so, the co-tidal lines should run across the North Sea roughly in an east-west direction.

Instead, what Whewell found was curious indeed. In the southern-most portion of the North Sea (called at that time the German Ocean) between East Anglia and the Netherlands, high water progressed south-ward along the East Anglia coast. But then it swung around to the east, moved northeastward along the opposite Dutch shore, and then back to East Anglia after twelve hours. In the more northern part of the North Sea, between northern England and Denmark, there was a larger but similar pattern. High tide progressed southward along the British coast, swept around to the east along the German and Danish coasts, and re-turned to the coast of northern England.

As Whewell wrote, "It appears that we may best combine all the facts into a consistent scheme by dividing this ocean into two *rotatory* systems of tide-waves." His resulting map showed the co-tidal lines radiating out from a point in the centre of each system, a bit like the spokes of a wheel, with the systems rotating in a counter-clockwise direction. (See Figure 8.1 for a modern map of tides in the North Sea.) And if there were such circular tidal motions around the North Sea, it implied that there might be points in the middle where there was little or no rise and fall of the tide at all. These "points of no tide" eventually came to be called amphid-romic points, and the rotating systems were called amphidromic systems. Later study showed that there was yet a third amphidromic system in the North Sea, with its central point close to southwestern Norway.

It was not easy with the technology of the nineteenth century to gauge the rise and fall of tides in the middle of a sea or ocean basin, un-less there happened to be a convenient island from which to take the measurements. But Whewell solicited help from the Royal Navy. In 1840, Captain William Hewett took his survey ship *Fairy* out toward the middle of the southernmost of the North Sea amphidromic system

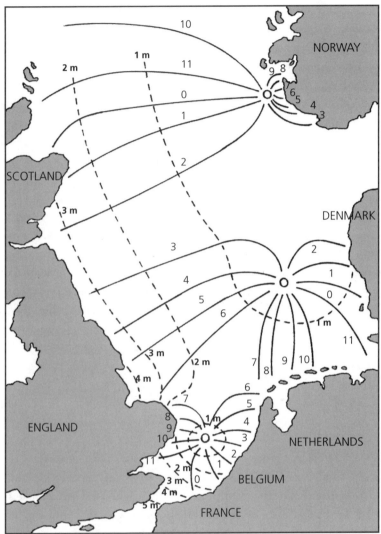

Figure 8.1: Amphidromic systems in the North Sea. Figures on the co-tidal (heavy) lines denote the time of high water in hours after the Moon has passed the Greenwich meridian. Figures on the co-range (broken) lines denote the mean tidal ranges in metres.

and anchored at a spot suggested by Whewell. He ran out cables in several directions to hold the ship steady over a shallow place with 21.5 fathoms (39 metres, or 129 feet) of water. Then depth measurements were taken over nearly two days as the tide turned repeatedly. As expected, the rise and fall remained within a narrow maximum range, only about half a metre (one and a half feet), or far less than the tidal range on the shorelines to the west, south, or east. Hewett and *Fairy* were lost with all hands in a fierce North Sea storm later that year, but Whewell had been vindicated.

Subsequent study has shown that similar, though much larger, amphidromic systems exist in all the world's major ocean basins. One of them, for example, rotates counter-clockwise around almost the entire North Atlantic. In the southern hemisphere, for reasons that will soon be apparent, they generally rotate the opposite way, in a clockwise direction.

Tidal science has also added co-range lines to the co-tidal lines to create complete tidal maps. The co-range lines, which look like concentric circles or partially closed loops, connect the dots of places with equal mean tidal ranges. At the centre of an amphidromic system, the range is zero. The farther out one goes from that centre toward the edge of an ocean basin, the larger the tidal range. That is why, as explorers like Dampier discovered (and as was reported from places like Bermuda), tides are generally small at islands well out from continental shores, and they become larger as the observer on a ship sails into shallower continental waters.

The Coriolis Effect

Whewell's discovery of amphidromic systems was just one of a series of major advances in tidal science that were made between the mid-eighteenth and mid-nineteenth centuries. Much of the pioneering work was done by scientists in France and elsewhere on the Continent. The outcome was to supersede Newton's equilibrium theory of the tides and replace it with what is known as the dynamic theory of tides. This was based largely on evolving mathematical analysis of the behaviour of waves, and it is also (and more descriptively) called the progressive wave theory of tides.

Newton's explanation of tides is called the equilibrium theory because it assumes that the tide-generating forces (resulting from centrifugal force and the gravity of the Moon and Sun) cause the ocean's waters to heap up. This creates the tidal bulges (as shown in Chapter 3), but the theory implies that the ocean's water can bulge only so far. At some point, a balance has to be reached between the forces generating the bulging effect and the pressure caused by the weight of the raised water to return the water to its original position. In other words, the tidal bulges represent an equilibrium situation.

As we have seen, Newton's theory is very useful in explaining how the relative positions and gravitational strength of the Moon and Sun can produce a difference between spring and neap tides. It also accounts for how, when the Moon's orbit takes it well north or south of the equator, a diurnal inequality can arise. But it fails to explain real-world observations on several other scores.

According to Newton's theory, for example — and here we will ignore the lesser influence of the Sun — an observer would always expect to experience a high tide at the time of lunar transit, i.e. when the Moon passes over the meridian (north-south line) along which the observer is standing. For the northern hemisphere, this would be when the Moon is in the sky due south of the observer (or else due north but below the horizon on the opposite side of the Earth). But actual observations show that, depending on the location, high water can come at any time before or after the Moon passes the meridian. That interval came to be called the establishment of the particular port.

Seen through the critical lens of the progressive wave theory, the real world behaves differently from Newton's model for a number of reasons. One is that he assumed the tidal bulge facing the Moon was free to sweep around the Earth as the Earth rotated. The bulge would always remain, therefore, directly "under" the Moon. To do this at the equator, however, it would have to move at about 1,600 kilometres per hour (1,000 miles per hour). But Newton's tidal bulge is actually a very long wave, with a high point (or crest) located beneath the Moon and a low point (or trough) located one-quarter of the way around the world. As waves were studied in the eighteenth century, theory showed that such a wave could only sweep around the world at that speed if the world ocean were at least twenty-two kilometres (more than thirteen miles) deep. If the ocean were much shallower — which it is, averaging closer to five kilometres (three miles) deep in most of the abyssal plains between the continents — the wave could not move nearly that fast.

Consider a tsunami, for example, which is also a very long wave.

When a large earthquake occurs in the Pacific Ocean, where typical water depth is around 4 kilometres (2.5 miles), the tsunami it causes travels at a speed of about 740 kilometres per hour (460 miles per hour). And that is also roughly the speed at which the tide wave propagates northwestward along the coast of California, Washington State, and British Columbia. Therefore, a real tide wave could never "keep up" with the Moon as the Earth rotates but would lag behind the theoretical bulge. And how much it lagged would depend on such things as the sea's depth and the latitude being considered.

Another huge limitation of the equilibrium theory is that, even if the ocean were deep enough, the continental land masses would block the movement of the tidal bulges. A long tide wave could not simply propagate around the world. It would be deflected by the continents. Then the configuration of the deep ocean basins and their continental shorelines would further channel the flow of that wave.

A lag in the tide wave could also be caused by friction as the turbulent flow of tidal currents moves along rugged or indented coasts, and in shallow seas, where the water has to move over a highly uneven sea bottom.

Finally, the Coriolis effect, which results from the Earth's rotation and its spherical shape, deflects any flow of water to the right (clockwise) in the northern hemisphere and to the left (counter-clockwise) in the southern hemisphere. To understand this is a bit tricky, so it may help to begin with an analogy from simple physics. Imagine that you had a gigantic cannon that could fire artillery shells over distances of thousands of kilometres. And suppose that you stationed this cannon on the equator and aimed it due north, say along the prime meridian, the longitude line that runs through Greenwich, England. If the shell were fired with enough power, it might land in the Arctic. But where it landed would be slightly to the east of the prime meridian.

Why this is so requires some explanation.

The Earth rotates in an eastward direction. At the equator, the circumference of the Earth is roughly 38,000 kilometres (24,000 miles), and our planet rotates once every twenty-four hours. This means that any point on the Earth's surface at the equator is moving eastward at a speed of around 1,600 kilometres per hour (1,000 miles per hour). Of course, we do not feel this when standing on the Earth. But imagine the effect of this movement on an artillery shell fired from a cannon. The kick of the explosion sends the shell out of the cannon in the direction that it was aimed, due north, and at a speed that depends on the power of the explosion, friction, air resistance, and other factors. Physicists would speak of a vector force being imparted to the shell, and that force is usually represented in diagrams as an arrow (denoting direction) with a variable length (representing the velocity imparted to the projectile). That vector would point due north, and its length would represent a speed of many thousands of kilometres per hour. (Giant German cannons in the First World War fired shells at about 6,000 kilometres per hour [3,700 miles per hour]. If our imaginary artillery shell exceeded roughly 40,000 kilometres per hour [25,000 miles per hour], the shell would escape the Earth's gravity entirely.) So the shell flies off in a northerly direction, arcs above the lower atmosphere, and eventually lands in the Arctic.

However, the cannon itself is also moving eastward at 1,600 kilometres per hour (1,000 miles per hour), the speed of the Earth's rotation at the equator. At the moment of firing, this movement gives the shell leaving the cannon an additional push. In other words, the Earth's rotation imparts a smaller second vector force to the shell, one that points eastward at a speed of 1,600 kilometres per hour (1,000 miles per hour). These two vector forces (one northerly, the other easterly) have to be combined to determine the effective resulting speed and direction of the projectile, which would be slightly to the east of due north.

Now, imagine for a moment that the Earth were not a sphere but a cylinder, with all the longitude lines running parallel to each other. In that case, the circumference of the Earth in the Arctic would be the same as at the equator, and any point on the Earth's surface in the Arctic would be rotating eastward at 1,600 kilometres per hour (1,000 miles per hour), just as at the equator. When the artillery shell reached the Arctic and landed there, the Earth would have rotated eastward just the right distance for the shell to land on the same longitude line, the prime meridian, from which it was fired. To observers on Earth, it would look as though the shell had followed a perfect path northward, and we would never notice that there had been an eastward vector at all.

But the Earth is not a cylinder. It is a sphere, and the longitude lines converge at the poles. In a sense, the Earth tapers, getting narrower at the poles. Its circumference in the Arctic (along any line of latitude) is much less than the circumference at the equator. Yet a point in the Arctic takes twenty-four hours to go around in each daily rotation, just as at the equator. This means that any point on the Earth's surface in the Arctic is rotating much more slowly eastward than the point on the equator from which the artillery shell was fired. So the eastward vector of the artillery shell will be *greater* than the eastward movement of any surface point in the Arctic. The Earth will not have rotated enough for the shell to land on the prime meridian. It will land somewhere to the east. (This is shown in Figure 8.2.)

What is true for an artillery shell hurtling high above the Earth also holds for the movement of water on the surface of the ocean, including any waves, such as a tide wave. This is because, like the artillery shell flying through space, that water is not effectively "attached" to the solid surface of the Earth itself, which is far below in the abyssal depths. Water in motion will be deflected slightly to the right (or clockwise) in the northern hemisphere, and this deflection had to be incorporated into the dynamic or progressive wave theory of the tides.

It may seem paradoxical, therefore, that an amphidromic system in the northern hemisphere rotates in a counter-clockwise direction. To understand why this is the case, picture a long wave that propagates northward and enters a large box-shaped bay that is enclosed on three sides and open only to the south. As the crest of the wave enters the bay, the Coriolis effect deflects it to the right, forcing the water to run up a bit higher along the eastern shore of the bay than in the middle. In other words, the water becomes somewhat sloped, with the higher side to the right. When this crest of raised water, moving rapidly northward along the eastern shore, reaches the back of the bay, at the northeastern "corner" of the box, where can it go? The contours of the bay and the pressure of the moving sloped water force it to swing around to the left, or westward. The Coriolis effect continues to deflect the wave to the right, so the same thing happens when the wave crest reaches the northwestern corner of the bay. Again, it is forced to swing around to the left, in this case southward. The result is that the wave tends to rotate around the bay in a counter-clockwise direction. And, in fact, the orientation of the bay — whether the "open" side faces north, south, east, or west — does not matter. In the northern hemisphere, amphidromic systems have a counter-clockwise rotation.

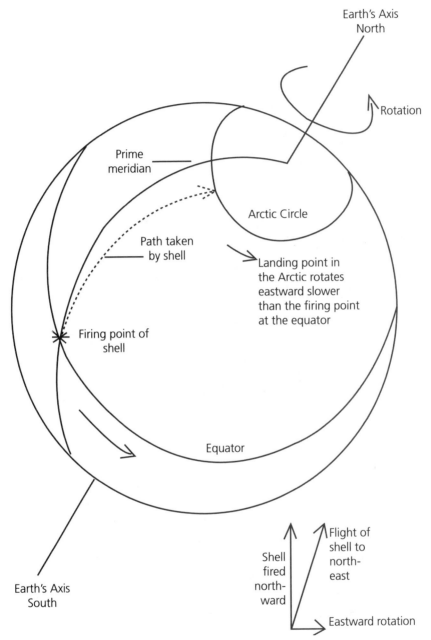

Earth's Axis
North

Rotation

Prime
meridian

Arctic Circle

Path taken
by shell

Landing point in
the Arctic rotates
eastward slower
than the firing point
at the equator

Firing point of
shell

Equator

Earth's Axis
South

Shell
fired
north-
ward

Flight of
shell to
north-
east

Eastward rotation

Figure 8.2: Illustration of the Coriolis effect.

Wave Interference

All of the shortcomings of the equilibrium theory came to be well recognized during the mid-to-late eighteenth century. In its place emerged the progressive wave theory, a somewhat more realistic, but also far more complex, view of how fluids behave under conditions that more closely resemble those that affect actual tides in the oceans on Earth. The emphasis was on understanding waves in general and how different waves interact with each other. The particular focus was on long ocean waves and how they are created by the rhythmic, regularly repeated impulses of the tide-generating forces.

Waves come in many shapes, sizes, and types, and the ways they can interact are complicated. But any wave — whether an ocean wave, a sound wave, or an electromagnetic wave such as a radio or light wave — is a periodic disturbance in a medium or in space. If it is a progressive wave, such as a wave propagating across the ocean, energy is transferred horizontally — in this case, across the sea's surface — by the vibration of the molecules. The surface of the water rises and falls as the wave passes, but the individual molecules of water do not, on average, move very far horizontally. (Actually, they circulate in tight little ellipses, which transfers energy from one molecule of water to the next, but then they return to their original positions.)

The periodic nature of a wave simply means that the highs and lows are repeated at a regular time interval, which is the wave's period. (The number of repeated highs or lows in any arbitrarily chosen span of time — per second in the case of radio waves — is the wave's frequency. The distance between two successive highs, or two lows, is the wavelength.) Where the tides are concerned, for example, a semi-diurnal tide wave responding to the Moon's gravity and the Earth's rotation would have a period (commonly called the lunitidal interval) of about twelve hours and twenty-five minutes.

A wave moving across the surface of a body of water may meet another wave coming from a different direction. In that case, they interact. The highs and lows (or the amplitudes of the waves) become superimposed. Depending on their wavelengths and the timing, the amplitude of one wave crest may be added to that of another, creating a higher resulting crest. Or the trough of one wave might be superimposed on the crest of the second wave, resulting in a reduced crest. This is called wave interference. In the actual ocean, waves coming from different directions (or waves being generated by different forces, such as storm winds in different and distant parts of the sea) are not likely to be synchronized with one another. Where they meet, their amplitudes will be superimposed without a regular pattern, leading to a chaotic set of local waves that toss and hurl themselves in every direction.

However, an irregular wave pattern, or wave interference, can result even in the case of waves that have been generated by similar forces and are moving in the same direction. Imagine that you dropped a pebble into water at the edge of a pond, and it caused a wave — a series of crests and troughs shaped like a sine wave — to move out across the surface. Then you quickly drop another pebble. Another series of crests and troughs moves out in the same direction. It could be that you just happen to synchronize the dropping of the second pebble so that the wave it produces is exactly timed to reinforce the first wave. Its crests would match the crests of the first wave. The resulting combined crests would be higher and the troughs lower. In that case, we say the two waves are in phase with each other. Their periods match, and in addition, as it moves, each crest passes any point along the surface at exactly the same time.

More likely, though, the timing of the second pebble will be at least somewhat off. The resulting waves will not be in phase with each other. Their crests and troughs will not match up as they move horizontally. Instead, the crest of one might be just a bit ahead of the crest of the other. In this case, the resulting combined rise and fall of the water's surface will not have the neat and symmetrical look of the typical sine wave. The pattern will be more complex, and it might be more irregular still if the waves have greatly differing amplitudes. For example, what if the second pebble dropped is much larger than the first? The amplitude of its wave would be larger as well.

Eighteenth- and ninteenth-century scientists began applying wave theory to their analysis of tides. They saw, for example, that if the gravitational effect of the Moon was twice that of the Sun, this was analogous to a situation where each of those heavenly bodies set up a tide wave, but one of those tide waves had twice the amplitude of the other. The lunitidal interval (the period between successive lunar "bulges") is twelve hours and twenty-five minutes, but the interval between successive impulses from the Sun's gravity is exactly twelve hours (one half a rotation of the Earth). This meant that the lunar tide wave and the solar tide wave would not be in phase with each other. Their crests would drift farther apart each day, generating an ever-changing combined tidal amplitude over the course of a fortnightly cycle from neap to spring tides and back again. (This is just what Newton's simpler theory predicts as well.)

Tidal scientists also found that it was useful to distinguish between free waves and forced ones. A free wave, such as that produced by dropping a stone into a pond, is one caused by a singular event or energetic impulse. As a free wave propagates outward across the water, its amplitude gradually diminishes as its energy is dissipated. An analogy would be an adult giving a single push to a child sitting on a swing. The child would swing back and forth, but air resistance would cause the arcs to become smaller each time.

A forced wave, by contrast, is one that keeps receiving energetic boosts. If you slap the water in a bathtub, you will make waves. If you keep slapping with just the right timing, those waves may get bigger and even slosh over the edge of the tub. Giving a well-timed extra push to the child on the swing each time it comes back to the adult would be an analogous situation. The arc of the swing will get higher. Because the Moon and Sun act on the ocean rhythmically, that is, with regular and repeated timing, tides are more like forced waves than like free ones. Instead of petering out over time, tide waves are reinforced with every rotation of the Earth.

Wave theory, combined with the sophisticated mathematics (especially calculus) that was available by the mid-eighteenth century, encouraged the attempt to understand the tides in new ways. Some, such as Joseph Louis Lagrange of France, studied the speed of the tide wave in the sea as it was affected by friction, the inertia of water, and the depth of the ocean. George Biddle Airy, Britain's Astronomer Royal, analyzed the behaviour of waves in canals and other enclosed bodies of water. He found that if the water depth and wavelength were just right, waves entering the open end of a bay might be reflected by the opposite closed end and bounce back, creating standing waves that moved up and down, rising and falling at fixed locations, but did not progress horizontally.

The brilliant French mathematician and astronomer Pierre Simon, the Marquis de Laplace, applied calculus to create a grand synthesis of all the celestial mechanics of the solar system, such as planetary orbits. This included the hydrodynamics of Earth's oceans. Laplace saw the tide wave as a forced wave that was set up by the tide-generating forces and had a period that was determined by the combined rhythm of those forces. He also introduced a highly innovative way of looking at the tide-generating forces and how the Earth's seas respond to them, but that's getting ahead of our tale.

Shortcomings of the Progressive Wave Theory

By the early decades of the nineteenth century, the prevailing concept was that tides were progressive waves that attempted to move westward around the world. Since the continents blocked their path, the only place where they could actually make the full circuit was in the Southern Ocean, although even there, Drake Passage between Antarctica and Cape Horn is somewhat constricted. Most tidal scientists assumed that, once past South America and moving westward, the tide wave would then propagate its disturbance northward in succession into the Pacific, Indian, and Atlantic Oceans.

If true, this implied that the same tide should arrive at places on both sides of the North Atlantic Ocean, for example, at roughly the same time. Which brings us back to William Whewell, who had proposed that simultaneous observations of the tides should be made in numerous places and that the progression of the tide wave along shorelines and across oceans should be mapped by the use of co-tidal lines. In 1833, two years before the international tidal experiment, Whewell wrote a preliminary paper about his proposed methodology. This showed hypothetical co-tidal lines progressing northward from the South Atlantic into the North Atlantic and running across the North Atlantic in an east-west direction.

Once Whewell had the results from the international experiment, however, he realized that something had to be wrong with this initial assumption of a progressive wave. As he wrote, the thinking until then had been that a "tide-wave [travels] to all shores in succession," which means that "the tide on the shores of America had been considered as identical with the tide on the coasts of Spain and Portugal, which occurs about the same moment; nor does it appear easy to imagine the form of the tide-waves so that this shall not be the case." But now there was a problem. Although tides on both sides of the North Atlantic are mainly semi-diurnal and fairly regular, sometimes they display a very significant diurnal inequality: one high or low tide will be considerably higher or lower than the following one. What Whewell found was that "the tides on these two sides of the Atlantic cannot be identical in all respects; for on the 9th, 10th and 11th of June [1835], when the diurnal inequality was great in America, it was nothing in the West of Europe; and on the 18th and 19th, when this inequality had vanished in America, it was great in Europe."

Whewell was also forced to take note of observations made by the great astronomer Sir John Herschel at the Cape of Good Hope during the same month as the experiment. They showed that the diurnal inequality there occurred at the same time as it did in Spain and Portugal. However, if the tide were progressing northward up the Atlantic from the Southern Ocean, its supposed source, this inequality should not appear until significantly later in Spain and Portugal, probably by at least a day or so.

Perhaps the most damning testimony against the progressive wave theory came from Captain Robert Fitzroy, the officer who commanded the survey ship HMS *Beagle* during a famous world-circling voyage — famous mainly because Charles Darwin went along as a young naturalist — who had also conducted an earlier surveying expedition on the *Beagle*. The carefully gathered data from this and other survey voyages on behalf of the Hydrographic Office was used by Beaufort's draughtsmen to produce the Admiralty Charts issued to British sea captains, which were highly regarded for their reliable accuracy by foreigners as well. Generations of British naval cadets were told to "Trust in God and the Admiralty Chart."

As part of the routine tasks assigned to Fitzroy for his voyages, he was ordered to observe the tides carefully. As he told the Royal Geographical Society upon his return to England, "beginning with the right or southern bank of the wide river Plata [Argentina], every mile of the coast thence to Cape Horn was closely surveyed." His observations on tides were eventually incorporated (along with other non-cartographic data such as information on winds and currents) into supplementary publications of the Hydrographic Office, such as the official *Sailing Directions* and *Nautical Magazine*.

During Fitzroy's circumnavigation, Beaufort informed him of Whewell's very early paper about co-tidal lines and the prevailing ideas about hypothetical tide waves progressing northward in the Atlantic. Fitzroy arrived back in Britain in 1836, just as the results of the international tidal experiment and its negative implications for the progressive wave theory were being analyzed and debated. In one of his published narratives of those voyages, he "ventured to ask whether the supposition of Atlantic tides being principally caused by a great tide-wave coming from the Southern Ocean, is not a little difficult to reconcile with the facts that there is very little tide upon the coasts of Brazil, Ascension [Island], and Guinea [in West Africa], and that in the mouth of the great river Plata there is little or no tide?" Fitzroy also claimed that there was no noticeable progression of the tide northward along the coast of Africa from the Cape of Good Hope to the Congo River.

Whewell was correct in his claims about the rotational systems in the North Sea, but the Royal Navy measurements that found the point of no tide in the North Sea and proved him right were not carried out until 1840. So, during the first few years following the international experiment, Whewell did not imagine that the same kind of rotating pattern might apply on a much larger scale for tides in all the world's oceans. In fact, he was not at all sure how to explain the big difference between tides on opposite sides of the Atlantic.

Meanwhile, however, Fitzroy weighed into the discussion. He wrote, with perhaps a touch of false humility, that "it may appear presumptuous in a plain sailor attempting to offer an idea or two on the difficult subject of 'tides.'" Daring to presume, he made a suggestion that later decades of tidal studies would show to have been prescient. Most likely, Fitzroy wrote, each ocean has its own tide, which he did not view as a progressive wave. Rather, influenced by the repeated and regular pull of the Moon's gravity (and to a lesser degree the Sun's), the water in those ocean basins might oscillate back and forth from east to west. The times of high tide would depend, therefore, on the breadth and depth of each ocean, and the pattern would vary, for the same reasons, from one part of an ocean to another.

By the mid-1840s, the failings of the progressive wave theory had begun to sink in, and a new approach was clearly needed. Airy, the Astronomer Royal, had been theorizing about how waves behaved in long, narrow canals. As mentioned, he found that if the length and depth of such an enclosed body of water were just right, a standing wave could be created, just as the water in a bathtub can be made to rock back and forth and slosh up and down in a regular fashion. With a progressive wave, both the wave crest and trough move horizontally in a single direction. With a standing wave, however, at each end of the canal (or bathtub) the water level alternately rises and falls in a rocking motion, a bit like the ends of a seesaw. At a fixed location halfway along the canal (or bathtub) there is a node (like the fulcrum of a seesaw) where the water level neither rises nor falls.

Airy rejected the idea that tides were produced by a single great wave propagating into the various oceans in succession. He argued instead that they might be more like standing waves, that is, forced oscillations that are set in motion in each individual ocean by the tide-generating forces. By that time, tidal measurements were available not only from both sides of the North Atlantic, where the tidal range tended to be quite large (the Bay of Fundy in Nova Scotia and the Severn Estuary across the Atlantic in Britain have the largest tidal ranges in the world), but also from mid-ocean islands, such as Bermuda and the Azores, which had much smaller ranges. This led Airy to suggest that the Atlantic tide was a stationary oscillation that had a nodal line running north and south up the middle of the Atlantic. To each side of it, the tide alternately rose and fell, much as Fitzroy had envisaged it. (Fitzroy, however, attributed the small tides at mid-ocean islands to interference between different and contending tidal oscillations.)

Whewell found yet another way of accounting for the same observations. Additional tidal observations that reached him after the 1835 experiment led him to believe that tides in the North Atlantic behaved much like those in the North Sea. He already realized that he could not draw co-tidal lines crossing the ocean from east to west. Instead, they seemed to sweep around the shores of the North Atlantic and meet the shoreline at an oblique angle. And since the tidal range at the mid-ocean islands was so small, just as at the points of no tide in the North Sea, Whewell thought the tide in the North Atlantic might also be revolving around a fixed centre. He proposed outfitting a naval voyage devoted solely to studying this question, just as an earlier British cruise had been sent out to study the Earth's magnetism, but the idea was not taken up by the Admiralty. Years later he tried again, suggesting a joint project with the U.S. Coastal Survey, but this, too, came to naught, and Whewell did not live to see his concept vindicated.

Instead, Airy's ideas of a cross-ocean oscillation prevailed for many years, although it, too, was eventually superseded. It was not until the early twentieth century that Whewell's basic scheme of a rotating amphidromic system in the North Atlantic (and elsewhere) was revived and, with some modifications, shown to be essentially correct. And the scientist who proved him right was an American.

The Great Wheel of Life

High Water and Coastal Biology

I huddled in the cold March drizzle with ten Nisga'a Indians on the deck of their gillnet fishboat. We were at anchor on the lower reaches of the Nass River in northwestern British Columbia, just east of the Alaska Panhandle. Within shouting distance were ten or more aluminum skiffs, flat-bottomed punts, and sleek outboard runabouts — and another two dozen fishermen. The Nisga'a waited, as they have done each spring for thousands of years, for millions of small eulachon (also spelled "ooli-chan") to arrive on the incoming tide. In the distance, clouds of seagulls advanced slowly upriver with the flooding tide, heralding the approach of the fish.

Suddenly the gulls were upon us. Their shrill cries filled the air as they whirled and dove for the fish. I ducked out of the way as the men jumped to their tasks. With deft teamwork, pairs of boats played out long, cone-shaped nets in the silty gray water, then spread them between stout hemlock poles that had been driven into the shallow river bottom. Within minutes, the eulachon were swimming into the nets.

On the nearest cluster of boats, two men used a long wooden hook to pull a tapered funnel of fine mesh, filled with fish, to the surface. Wielding a shorter hook, another man worked a manageable bulge of fish down to the end of the net. Then two others lifted it over the gunwale of their empty skiff and released a slip knot. Hundreds of slim, silvery fish, flashing and jumping, slid out at their feet. Working quickly, the men kept squeezing fish along the net, like toothpaste in a tube, as masses of them continued to surge into the wider submerged mouth of the net.

As the first full skiffs wallowed to the nearby riverbank to unload, empty boats darted into position at the nets. The roar of outboard motors echoed off the nearby hillside. For three hours, until slack tide, the Nisga'a raced to harvest their most precious food resource.

The eulachon, *Thaleichthys pacificus*, is a smelt so rich in oil it is also called the candlefish, because when dried and fitted with a wick it was burned by Natives for lighting in the days before electricity. From mid-March through the end of April, in runs lasting a week or two, eulachon swarm with the flooding tide into the large mainland rivers of the Pacific coast from northern California to the Bering Sea. Since long before living memory the fish has meant life itself to Natives on the B.C. coast. "They were our first fresh harvest of the year," the late Nisga'a elder Henry McKay told me when I visited the Nass. "They came when our people were at the end of their winter food supply of dried fish, seaweed, roots, and berries. So we named it the Saviour Fish."

In good years the catch was enormous. A century ago each family took five to ten tons for its own use every year. After gorging on fresh fish and smoking or drying a portion, they rendered the bulk into a rich, concentrated oil they called grease. Kept cool, it could be stored for at least five years.

Northwest coast peoples relished this high-energy grease with everything, even berries and desserts. Many of their smoked and dried foods could hardly be swallowed without the help of grease or other fish and animal oils. Grease was also used as fuel, lubricant, and medicine. In the 1870s the street lamps at Metlakatla, a Native village near Prince Rupert, burned eulachon oil. The Nisga'a still use grease as a laxative and for burns, chapped hands, and colds.

Tribes that were located on good eulachon rivers were fortunate. The Nass, where the tidal range can exceed six metres (about twenty feet) and the tidal currents sweep in and out of the river at two to three knots, had by far the largest eulachon fishery of all B.C. rivers. "It meant survival, power, and wealth to the Nisga'a," said Alan Moore, who has researched his people's history. "We controlled it." Others had to acquire the prized delicacy through trade. In the words of one early observer, George Chismore, "By canoe it travels to Sitka on the North and Puget Sound on the South, as well as up all the navigable rivers. Inland, borne upon the backs of men, it goes, no whiteman knows how far; certainly to the Arctic Slope, traded from tribe to tribe, and becoming more costly the farther it gets from its source."

Tribes from the interior trekked through the coastal mountains to trade rabbit, marmot, and beaver skins for the grease, making their way along narrow, well-worn "grease trails." It was along such a route to Bella Coola that explorer Alexander MacKenzie reached the Pacific by land in 1793. The grease trail to the Nass was such a major trade route that the Hudson's Bay Company established Fort Simpson at the mouth of the Nass in 1831 in an effort to capture a share of that commerce.

Even in recent decades on the Nass River, the Nisga'a have caught roughly two hundred tons of fish most years, enough for every interested family to enjoy a big feast of fresh eulachon and lay in a supply of the rendered grease that is their favourite condiment. The annual run remains a defining event in their way of life, as I discovered when I drove to the isolated Nass Valley and hitched a boat ride down the river. We motored by under the fixed gaze of countless bald eagles, perched statue-like in tall trees, watching for fish. On other branches, strung out like black picket fences, were platoons of ravens. Twenty-two kilometres (fourteen miles) above the river's mouth we came to the ramshackle seasonal settlement at Fishery Bay. Accessible only by boat, it was used by Nisga'a from the entire valley for only six weeks each year, during the eulachon harvest.

I counted five camps on the bay, each run by a company of relatives and friends who pooled their labour and equipment and shared in the catch. The camps were no-nonsense hamlets of weathered wood frame houses and fish processing sheds. Boats flitted back and forth between the nets and riverside floats. Chainsaws droned. Kids in gumboots romped along the muddy river bank. Woodsmoke drifted from every chimney.

I went ashore to look in on one of the camps and met Alan Moore, who showed me a log-walled bin the size and depth of a small swimming pool. It was already half full of fish. Two teenage boys lugged a heavy box of eulachon up a planked walkway from the river and dumped them in. "We'll work every tide until it's full," said Moore. Then some forty tons of fish would be allowed to "ripen" for one to three weeks. This releases the oil, making it easier to render. The stench can be offensive to outsiders, but camp workers don't mind. "You get used to it," Moore chuckled, "just like the white man's polluted cities."

A woman was minding the camp and keeping an eye on the wood fires and the three girls and half-dozen teenage boys of the shore crew. I caught her up to her elbows in blood as she sliced long strips of fat from a huge, meaty carcass. It was a sea lion that had followed the eulachon upriver and been shot by one of the fishing crew. Her young daughter and niece were busy hanging the blubbery strings in the smoke house.

After landing more than two tons of fish on a favourable tide, the men came in from the river to warm up, relax, and grab a bite to eat in their company's house. Alongside a tray of sandwiches was a platter of dry, thinly sliced smoked salmon and a jar of clear oil. I tore off a piece of the jerky-like orange flesh, dipped it in the oil, and savoured my first taste of traditional Indian salmon and grease. The oil added a pungent flavour and offset the saltiness of the chewy fish.

At dusk the men headed back out into the blustery weather, but the evening tide was a weak one that brought few fish. After two hours the men straggled back in, cold and disappointed. They peeled off their rain gear and tucked into a late dinner: heaping platters of juicy red sea lion steaks, which tasted like lean beef.

After dinner the youngsters turned in, while the adults stayed up until midnight playing word games and swapping stories. They told of hair-raising escapes while fishing the unstable river ice, of scrambling up the nearby crags to hunt mountain goats, of grizzlies coming into the camp and knocking over barrels of grease. Then they drifted off to their sleeping bags, only to drag themselves out again the next day to fish another large tide. But the weather took pity on them. Brilliant sunlight traced the snowy ramparts of the Coast Range against a deep blue northern sky. Out on the boats, the men worked their nets cheerfully. A few more days of good fishing — a few more strong flood tides, like this one — and the bins would be full. As it had done for millennia, the swift tidal flow was delivering a biological bonanza.

Tidal Action Delivers a Wealth of Nutrients

It is hardly coincidental that one of the richest Native fisheries on the northwest coast is located on a river with a large macrotidal range and powerful tidal currents. The same pattern holds true in many places around the world. Large tides and powerful currents generate a mixing of cold, deep ocean waters with warmer surface waters. The colder bottom waters are full of plankton and tend to be rich in dissolved oxygen and organic nutrients. When thrust to the surface by the turbulence of vigorous tidal mixing, these provide sustenance for a wide range of diverse marine species: not only fishes, but also invertebrates, birds, and sea mammals.

Strong tidal currents combine with other factors to create particularly food-rich ecosystems in warm-water areas as well, and especially in places with brackish water, where fresh water and ocean water mix. One study of estuaries on the coast of Georgia (home of those large oyster reefs) found that among the factors leading to high primary productivity — that is, the abundant growth of organisms at the bottom of the food chain, such as phytoplankton and diatoms — strong tidal currents were the number one contributor.

In fact, the word *estuary* derives from the Latin *aestus*, meaning "tide." As Wikipedia notes, "The key feature of an estuary is that it is a mixing place for sea water and a stream or river to supply fresh water. A tide is a necessary component to maintain a dynamic relationship between the two waters. Though something in the nature of an estuary can exist in a non-tidal sea, such areas go by names such as *lagoon*, *étang* or *laguna*. In non-tidal seas, the rivers naturally form deltas rather than estuaries."

As we have seen (in Chapter 5), on temperate-zone coastlines with large tides and strong currents, typical estuaries are characterized by large areas of lush salt marsh. These are dominated by marsh grasses (notably *Spartina alterniflora*, or saltmarsh cordgrass, and *Juncus roemerianus*, or black needlerush). In the course of an annual cycle of growth and decay, the grasses contribute enormous quantities of minute organic detritus to the ecosystem; this, in turn, becomes food for shrimp, crabs, small fishes, and other organisms higher up the food chain.

A classic study in the late 1960s by John and Mildred Teal of the eastern coast of the United States presented quantitative comparisons between the primary productivity of estuaries and other ecosystems.

The unit measured was tons of dry weight matter produced annually by each type of ecosystem for an area of 1 acre (about 0.4 hectares). For dry agriculture, such as wheat farming and forestry, productivity ranged from 0.3 to 1.5 tons. In moist (or irrigated) areas on land, it was 1.5 to 5 tons. Coastal ecosystems (other than estuaries) produced 1.0 to 1.5 tons, but only 0.3 in the open ocean. In estuaries, however, productivity ranged from 5 all the way up to 10 tons or more, making them vastly more productive than other ecosystems, and that productivity supported organisms extending right up the food chain.

As the Teals summarized it, "Estuaries in general and salt marshes in particular are unusually productive places. None of the common agricultures, except possibly rice and sugarcane production, comes close to producing as much potential animal food as do the salt marshes. The agricultural crops which approach this high figure are fertilized and cultivated at great expense. The marsh is fertilized and cultivated only by the tides."

Marshes are so rich "for several reasons, all of which are a result of the meeting of land and sea. The tides continually mix the waters and, by their rise and fall, water the plants. Harmful accumulations of waste products are diluted and removed. Nutrients are brought in continuous supply." They went on: "The plants can put energy into growth, which in another environment they might have to use for collecting nutrients. The meeting of fresh and salt waters tends to trap nutrients in the regions of such meeting and this concentration of nutrients promotes plant growth."

Finally, estuaries and salt marshes are uniquely rich places "because there is almost no time during the year, even in the north, when there is not some plant growth taking place. In the south, it is warm enough for the *Spartina* to grow all year.... But in the north, where the land plants and *Spartina* cease activity during the winter, the algae in the marshes continue to grow throughout the year." And both the *Spartina* and algae then become food for animals. "Of the bounteous ten tons of possible food per marsh acre, the insects feeding directly on the *Spartina* use only a small part, less than four-tenths of a ton. Most of the *Spartina* is left to die and be decomposed by bacteria."

This bacterial decay makes "about five tons of suitable detritus-algae food available every year from each acre of marsh. The detritus-algae eaters [such as certain snails, crabs, shrimps, and small fishes] eat about one-ninth of this food. Both groups of animals, the plant eaters and the detritus-algae eaters, use up most of the energy they get from their food and put the rest into the building of their bodies. The carnivores live on these animals, consuming about one hundred and fifty pounds [per acre]." Even adding all these types of consumption together, however, "we have accounted for only fifty-five percent of the energy that remained after we allowed for use by plants. Four and a half tons dry weight per acre of marsh production is left over." That remaining potential energy is not wasted, though. "The tide giveth and the tide taketh away. It is the tide that makes the high production possible and then removes [almost] half of it before the animals of the marsh get a chance to use it. But what is denied the animals of the marsh is given to the abundant animal life in the estuarine waters around the marsh."

If the surge of tides is so important, it is not surprising that changes to their ebb and flow can destroy the health of estuary ecosystems. As noted in the *Tidal Crossing Handbook* (published by a Massachusetts-based environmental organization), "Preserving the marsh's natural tidal range is the key to maintaining productivity. The cycle of inundation and drainage of the marsh is essential to the vitality of salt-marsh plant species." And as New Jersey biologist Paul Jivoff told a reporter for the *Washington Post* in 2003, "For a salt marsh to thrive, its grasses — and the aquatic life they sustain within their fronds — need open tidal channels and the flushing of nutrients in and out of the marsh. Tidal flow is a salt marsh's essential circulatory system." Restrictions on that flow are mainly due to human activities such as building dikes or roads, which is why industrial and urban development on the U.S. eastern seaboard has been so harmful to salt marshes.

Dolphins and Manatees

The tides and tidal currents of the U.S. Atlantic coast are relatively modest ones. Some species, however, thrive especially well in places with truly extreme tidal conditions. One of the most interesting is Commerson's dolphin (*Cephalorhynchus commersonii*), a small, blunt-headed, black-and-white toothed cetacean that never weighs more than 86 kilograms

(190 pounds) or exceeds 1.7 metres (5.5 feet) in length. It ranges from the Falkland Islands to the islands of extreme southwestern Chile to the South Shetland Islands near Antarctica. But it is mainly found along the shores of Patagonia and Tierra del Fuego and especially in the Strait of Magellan, where a study in 1984 estimated its population at thirty-four hundred.

Whereas some whales and other cetaceans avoid the turbulence of swift tidal channels, Commerson's dolphins *prefer* such places. They feed in areas with the largest tidal range, moving in toward shore with the flooding tide. And they frequent places with the strongest currents, such as at the eastern end of the Strait of Magellan and particularly the First and Second Narrows, where the current often runs at eight knots or more. One study of what was found in the stomachs of dead animals showed that they eat mainly tiny mysid shrimp (which are smaller than true shrimp, inhabit mainly shallow, intertidal waters, and hide in kelp beds), several species of small fin fishes, squid, and some algae (seaweeds).

They like to feed in the dense kelp forests that line much of the rocky shores of Tierra del Fuego, but also around artificial structures such as oil rigs and piers. And far from avoiding rough water, they positively frolic in it, often leaping into the air and riding breaking waves near shore and the bow waves of fast-moving ships and boats. A study of the dolphins living north of the Chilean end of the Strait of Magellan found that they gravitate toward places with rapid tidal flows, tide rips, and shallow water over sand banks at the entrances to fjords. They seem to use swift water flows to help acquire food, feeding along the edges of adjacent or contending currents. They also form cooperative groups of up to fifteen individuals to herd schools of fish against the shore.

By comparison, let's look at a much larger sea mammal, the manatee. Weighing between 450 and 900 kilograms (1,000 to 2,000 pounds), manatees are herbivores, closely related to the elephant, that like to graze in calm, shallow water with small tides and minimal tidal currents. One time, I was visiting friends on their sailboat, which was tied up at a dock on a river in Fort Lauderdale, Florida, where the tidal range is very small, usually less than one-third of a metre (one foot). Enjoying a quiet morning coffee out in the cockpit, I suddenly heard a loud splash behind me. By the time I turned to look, the smooth, gray, glistening back of a huge animal was just disappearing below the surface of the murky water. I had no idea what it could be, and my friends were amused at my shock, because they were already accustomed to having manatees nosing around their boat.

Manatees are also called sea cows. Even the name hints at the difference between their lumbering nature and that of the swift, agile dolphins. In fact, their slow and docile habits are their greatest vulnerability. Many manatees are injured or killed each year by motorboat propellers in coastal waters. Around the Florida coast and in the Caribbean waters, they prefer sheltered rivers with slow flowing water and access to fresh water. They also frequent coastal lagoons with abundant mangrove forests and estuaries lined with grasses. The contrast between the preferred tidal habitat of the manatee and Commerson's dolphin could not be greater.

A Hot Spot of Biodiversity

Our entire natural maritime world would be far poorer if it were not for the biological wealth contributed by places with large tides and/or powerful tidal currents. Many of them are localized hot spots of biodiversity and productivity. At the various tidal rapids on British Columbia's coast, for example, the upwellings of cold water bring nutrients to the surface that feed large populations of jellyfish and shrimp, which in turn attract vast schools of herring. The herring then become food for millions of larger predatory fish, such as salmon, as well as for flotillas of feeding gulls. One fishing guide found a wonderful metaphor for this swirling, cyclical, and interlocking food chain. He called it the "great wheel of life."

As on the Nass River, high-current areas also become magnets for the beasts highest on the food chain: seals, sea lions, dolphins, and whales. I once went salmon fishing at Active Pass, a tidal-torn narrows that separates Mayne and Galiano Islands, not far from where I live. The tide was running between the islands at five or six knots, creating lines of foaming froth and high standing waves. Poking out of those waves were the whiskered snouts of sleek, shiny California sea lions. Even with constant attention to the throttle and gearshift, we were barely able to keep our outboard runabout in a stationary position, which was important for trailing our hooks and lines into the tidal flow. But the sea lions, incredibly powerful swimmers, did not seem to be having any problems at all. They just stared at us in apparent amusement as they bobbed on the faces of wildly tossing, chaotic waves, eagerly anticipating a succulent meal of fresh coho or spring salmon.

Scientists who dive in these channels (which they do, for safety, mainly during slack tide) find them to be havens for a great diversity of sea life. One of the best-studied places in the world with swift tidal currents is Race Rocks, a tiny, windswept lighthouse island (with off-lying rocks and shoals) in the Strait of Juan de Fuca, southwest of Victoria, B.C. The thirty-kilometre (almost twenty-mile) wide strait is where the tide from the open Pacific surges into and out of the sheltered and constricted inner waters of Puget Sound and the Strait of Georgia. Race Rocks is a major feeding and haul-out area for some fourteen hundred sea lions and four hundred harbour seals. It is also an occasional mating site for huge northern elephant seals. But the more interesting story is what goes on beneath the surface of the turbulent waters.

In the early 1990s, I spoke to Garry Fletcher, a professor of biology at nearby Pearson College. Fletcher, who retired in 2004, was the key person involved in establishing an ecological reserve on and around Race Rocks in 1980 and managing it as a volunteer warden for many years. One of the first projects he oversaw was the installation of an underwater current meter, which was anchored to a huge concrete block on the sea bottom. The meter found that the tidal currents routinely run at six to six and a half knots past the rocks.

And the concrete block itself has provided revealing information. After ten years on the bottom, ten metres (thirty-three feet) below the low tide level, it had become thoroughly encrusted with marine life. "This year," Fletcher told me, "a student is studying the five [exposed] faces of the block and the life that's grown on it," which varies from one surface to another because of "different exposure to currents and to light." But the natural sea floor had not been neglected. Other student divers had been involved for many years in keeping track of changes over time in sea life at fourteen baseline monitoring stations. Students had also been conducting repeated underwater transects, recording what was living along a swath that was thirty metres (about one hundred feet) long, one metre (just over three feet) wide, and extended down to about ten metres (thirty-three feet).

"At the moment," he said, "we're finding lots of hydroid species. That's the anchored stage in the life cycle of jellyfish. Dr. Anita Brinckmann-Voss, who is an expert on these, has identified up to forty-five species of hydroids, a very rich diversity. This is typical of high-current areas." The hydroids — some of them look like long, very fine feathers, others more like stalks that are sprouting flowers or onion bulbs on their ends — do very well at Race Rocks, Fletcher added, because "they need a lot of food and areas that are not easily grazed by sea stars" and other predators. "Their food is plankton. The higher the current, the more plankton there is going by per minute, which makes more food available." And the plankton was being churned up by the currents. "At Race Rocks, water from the depths of the Strait brings an upwelling of nutrients from the open Pacific. It is a very rich zone, not only for hydroids but for other invertebrates, too."

Fletcher mentioned crabs, particularly blue clawed (lithode) crabs, which also thrive in some of the tidal narrows such as Skookumchuck Rapids, and several species of anemones. "The brooding anemone, *Epiactus prolifera*, only grows in high-current zones. At Race Rocks it's very abundant and in many colours." Anemones tend to cling to the vertical sides of rocks and channels. "Different sea anemones handle the velocities of the currents differently. Some are built short and compact, and they can handle the pounding surf zone" in shallow water. "Others are built so they can extend themselves out quite a way. *Metridium senile*, or plumose anemone, for example, is a plankton feeder. It can extend out a foot and a half [one-half metre]. It has a stalk and then a big mushroom of tentacles on top, and they'll snag the plankton. They never grow as solitary animals. They're always in clumps," which is an extra adaptation to strong currents. The form of such a colony "makes for a more laminar flow around it. And where you get a lot of flow along a wall, you'll get solid masses of plumose anemone."

The sea bottom at Race Rocks is also home to large scallops, which, along with crabs, shrimps, and rockfishes, provide food for the giant Pacific octopus (*Enteroctopus dofleini*). This amazing mollusk can weigh up to 270 kilograms (595 pounds) and have an arm spread of almost 10 metres (33 feet). The octopuses are then fed on, in turn, by seals, large lingcod, and dogfish, which are small sharks. The seals and sea lions are preyed upon by the unchallenged kings of the region's bestiary, the transient orcas, while the resident pods of orcas feed mainly on salmon and other fishes. The swift currents flushing in and out of the open Pacific with relatively large tides help to create an extremely rich web of natural life.

Hawaii and French Polynesia

In some situations, even quite small tides can have a powerful effect on currents and ocean mixing. Hawaii has only a microtidal range, usually only about two-thirds of a metre (a little over two feet), yet surprisingly swift tidal currents flow in the open ocean around the major islands.

I first heard about this from a former professional beach boy, Thomas Vinigas, who worked as the pool attendant at the Ilikai Hotel in Honolulu, where my wife and I were staying. In his free time, Thomas is the sparkplug of a competitive canoe club based near his home in Kailua. He serves as the captain, steersman, and overall strategist of a large outrigger canoe (with six paddlers) that participates in long inter-island races, such as from Molokai to Oahu, which can last seven hours or more. He has to be aware of the time and the changing tidal conditions, because if he runs up against a strongly flooding or ebbing tide going the wrong way, it can stop his canoe dead in the water. On the other hand, if he times it just right, that current can sweep his canoe along at what feels to the paddlers like almost breakneck speed.

"Just how fast are these currents?" I asked him. The next day, he brought me a sheaf of photocopies showing the speed and direction of the prevailing winds and currents around the Hawaiian Islands. Part of the material was charts of the shorelines of Oahu, with neat little wind and current diagrams that had arrows showing the percentage of time (which changed by season) that each vector has a particular strength. And there was a summary page that made the important distinction between the prevailing wind-driven current (the North Pacific Equatorial Current), which flows predominantly westward in the Hawaiian Islands, and tidal currents, which reverse themselves twice daily and can, therefore, run contrary to the wind-driven current. (This is the same distinction that Dampier made between "currents" and "tides," which for him included what we now call tidal currents.) "The ebb and flood currents reverse off most of the leeward coasts and off windward Kauai," the document went on. "The current velocity varies greatly" and can run at over one knot.

Over one knot sounded pretty fast for an open ocean current, especially one being generated by such small tides. After all, the powerful Gulf Stream flows at an average speed of less than two knots, and the infamous Agulhas Current off the east coast of South Africa has a speed of around three knots. So I was a little skeptical, and wondered how such small tides could produce such fast currents in unconstricted waters. I learned the answer by visiting oceanographer Mark Merrifield in his office at the University of Hawaii. Merrifield has done extensive research on so-called internal tides around the Hawaiian Islands. These differ from ordinary tides, which are long surface waves that have wavelengths in the thousands of kilometres and where the surface of the water rises and falls. Merrifield explained that internal tides are "huge undulations" in the density of deeper ocean waters caused as the ordinary tide wave propagates across and around an ocean basin. The wavelength of internal tides is much shorter, in the tens of kilometres. What is moving along horizontally in the case of an internal wave is the boundary between denser and less dense water. That boundary rises and falls in a regular wave-like pattern as the crests and troughs propagate, and then the wave reverses itself and propagates in the opposite direction twice a day.

In the case of Hawaii, the ordinary tide wave encounters a huge ridge — the island chain is essentially a mountain range as high as the Himalayas — coming up from the ocean floor. "Twice a day, the tide has a flow that goes back and forth across the ridge. And as it does so, it can do two things. It can either sort of squirt its way through the little passes" between the major islands, "or it goes up and over the seamount ridges. And when it does that, it generates waves in the ocean that we call internal tides." A vast quantity of water is forced rapidly up and over the ridge, which causes the density of that water to oscillate in wave-like fashion. And in terms of the energy generated by this repeated subsurface upheaval, Hawaii is "probably one of the most energetic sites for internal tides in the world."

These internal tides cause extensive mixing within the ocean, which evens out the differences in temperature, salinity, and density between different layers. And they also result in strong surface currents.

Merrifield and his colleagues were asked a few years ago by local canoe racers just what kinds of currents to expect in the channels between the islands. Well, he said, warming to the story, "it turns out that the main current is actually the surface expression of these waves," the internal tides. So, whereas ordinary tidal currents in the open ocean are quite weak, those produced by these internal waves are much stronger. The people requesting advice were "trying to understand how you might steer your canoe between Molokai and here [Oahu]," which meant one had to "take into account how these internal waves are generated. Because they're generated by the tides, they're somewhat predictable. So, we've come up with models of what that flow should be." And approximately how fast could the currents run, I asked. Merrifield confirmed the information from the documents Thomas Vinigas had shown me. "You can get flows up to a knot or even a knot and a half in some instances. And that's pretty fast for the ocean."

Where they are forced through the narrow entrance to a lagoon, even small tides can also produce relatively swift currents. And where this happens, those currents can sweep with them an abundance of fish, a situation not very different from the one created by the very large tides in places like B.C.'s Nass River. I got to witness this a couple of years ago at the island of Huahine, which is situated northwest of Tahiti in the Society Islands of French Polynesia. Near the scenic little village of Maeva, on Huahine's northern end, is a shallow lagoon. The islanders call it a lake, but it is actually a place where a narrow channel brings ocean water into a wide and shallow body of mainly salty water that lies between the main (volcanic) island and an outer line of linked coral barrier islands, or motus. And in that channel is a series of highly productive ancient stone fish traps that have been restored in recent years and are now in continual use.

Tahiti lies near an amphidromic point, which means that its tidal range is extremely small, rarely more than one-third of a metre (one foot). At Huahine the range is larger, but not by much. Te Tiare, the lovely and quite isolated resort where my wife and I stayed, had bungalows and a main building that were built right out over the lagoon. It was accessible only by boat, but the dock where the boat arrived did not need to have a ramp or other feature to accommodate the rise and fall of tides. Everything was simply built on concrete pilings. Over a couple of

Photo by Tom Koppel.

In the tidal channel at the island of Huahine in French Polynesia, stone traps are used to catch fish. It is an ancient practice that is still in use today.

days, I noted the lines left by wetness and marine growth on the pilings, which revealed the normal tidal range. It was certainly less than half a metre (one and a half feet).

When we visited the "lake" at Maeva, however, the flow of water into the lagoon was remarkably swift. The tide was flooding in at what I reckoned to be a speed of around two knots. And it was bringing with it a steady stream of lagoon fishes, which two or three islanders were busy harvesting. They did this with the help of fish traps that are made of stones piled on top of each other in the shallow channel. Essentially, these are stone walls that are high enough to break the surface even at high tide and are aligned to form long Vs. The walls are widely spaced on one end of the V and converge in the other direction at small, round enclosures. The fish are swept by the current into the wide opening, but when they reach the narrow end, escape is difficult or impossible. Perched over the enclosures are small, open-sided thatched huts, where the people tending the fish trap can store equipment and hang out, shaded from the intense tropical sun, while waiting for the tide to do its work.

During our visit, burly Native guys were working two of the five traps, using large nets to scoop the fish out of the stone enclosures. If they harvested more fish than they needed at the moment, there was also a wire mesh box immersed in the channel, where they could place fish and keep them alive. It was a brilliant system, and one that had been there for centuries. In fact, this set of traps had fallen into disuse until a decade or so ago, when it was surveyed and restored by archaeologist Yosihiko Sinoto of the Bishop Museum in Honolulu. The fellow driving my wife and me around the island and showing us the ancient sites was Paul Atallah, an American who had studied archaeology in Hawaii and worked with Sinoto on site restoration on many islands in French Polynesia. Atallah recognized one of the men working the traps, whom he had never seen doing so before. "He lost his job last year," Atallah said, when one of the few resorts on Huahine shut down. "So he may be catching fish to sell, but more likely he just needs the fish to help feed his family."

As on the Nass River with its large tides and traditional native fishery, the small (but regular and predictable) tides of Huahine were providing the islanders with a steady supply of high-quality food.

Sea and Sky in Harmony

Oscillating Waters and Modern Tidal Forecasting

One morning while visiting the Bay of Fundy, my wife and I got up early to take pictures at low water and see the dramatic spectacle of the world's largest tide going through its cycle. We were not disappointed. We drove to the small town of Hantsport, on the estuary of the Avon River, just upstream from where the Avon flows into the Minas Basin, the area with the bay's largest tidal range. Hantsport's public wharf is one of the best places to observe the tidal extremes. I had seen photos of the wharf at low tide, perched high and dry, with no water in view, on wooden pilings twelve metres (forty feet) tall. Other photos, taken at high tide, showed water lapping just below the tires of cars parked on the wharf.

When we arrived, the wharf's pilings were bone dry for their entire height. The nearest water, a few murky pools linked to the shallow river, was far away across broad mud flats. A local guy who was enjoying the morning sun out on the wharf told us that we would have to wait quite a while for the tide to come in. But when it did, we would have the chance to see a huge bulk carrier loading up on gypsum at the adjacent wharf of the Fundy Gypsum Company. We went off for a leisurely breakfast and returned a couple of hours later.

By then the entire scene had changed. The flooding tide had come in and was at least one-third of the way up the pilings at the public wharf, which was now full of people hanging out to fish and gab. And a giant ship was tied up at the nearby gypsum loading wharf. It had arrived only a half-hour earlier, riding empty, light, and high in the water. Huge hatches were open, and long conveyor belts were dumping the dusty white pulverized gypsum into the ship's holds. But the loading procedure had just begun. The ship's painted waterline was still about six metres (twenty feet) above the water of the estuary.

In fact, we learned from a representative of the gypsum company, the bulk carriers could only come in to load up while totally empty (when their draught was at its minimum) and on a flooding tide. And then loading had to be completed in about three hours, because once the ship was full, it sat so much deeper in the water. To avoid running aground, the loaded ship had to head out of the shallow estuary immediately, while the tide was still high. There was little margin for error.

The day we watched, all went perfectly to plan. In early afternoon, the loading was finished. The ship, full to the brim, rode right at its painted waterline. The crew cast off the lines and a tug helped to nudge it and guide it out toward the sea. The tide had risen over fifteen metres (about fifty feet) in six hours and was lapping at the very top of the wharf pilings. We had witnessed the flood cycle of the world's largest tides, and it is truly remarkable. Because the bay is so large — 250 kilometres (155 miles) long and about 60 kilometres (37 miles) wide — to raise its level that much means that the volume of water surging into the bay twice each day from the open Atlantic equals the combined flow of all the rivers on Earth.

The Natural Harmonics of Physical Systems

Why is the tidal range on the Bay of Fundy larger than anywhere else in the world? (Actually, recent measurements show that it is just about equalled by the tides on remote Ungava Bay in the Canadian Arctic, but the Fundy tides have been much more thoroughly measured and studied.) The answer requires an understanding of the third major advance in tidal theory (after Newton's equilibrium theory and the dynamic, or progressive wave, theory). This is harmonic analysis. It is a method based on the phenomenon of harmonics, which also includes the closely related concept of natural resonance between bodies of water on Earth and the astronomical tide-generating forces.

Objects and physical systems often have unique natural frequencies of vibration or oscillation. (For slowly oscillating systems, it is more convenient to speak of their natural periods — measured in seconds, minutes, hours, days, or even years — rather than their frequencies — measured in vibrations or cycles per second. But the principle involved is the same; period is simply the reciprocal of frequency.) These frequencies depend on the object's length, thickness or density, tension, and other factors. If a periodic force acts on the object at (or close to) the same frequency as the object's natural frequency of vibration, it will cause the object to vibrate at that frequency and to do so with a large amplitude of vibration or oscillation. This is called resonance.

A suspension bridge, for example, may have a natural period of oscillation, which makes it vulnerable. That is why a troop of soldiers will fall out of step when they march across one. If the rhythm of the soldiers' usual march step happens to coincide with that period, the bridge may begin to resonate with that rhythm. With each step of the soldiers, more energy is pumped into the system, causing the bridge to sway wildly or even to be destroyed. Something similar happened in 1940 to the Tacoma Narrows Bridge in Washington State, which developed resonance with repeated gusts of wind. Each swing got larger until the bridge snapped and crumbled.

A pendulum, too, has a natural period of oscillation. It does not swing back and forth with just any period, nor does its period of swing depend on how energetically it is pushed when set swinging. Each pendulum has a natural period of swing that depends exclusively on its length. (The amplitude of its swing — i.e., the length of the arc it makes — does depend on how energetically it is initially pushed.)

Some objects or physical systems have more than one natural period (or frequency) of oscillation, which means that there can be resonance with periodic forces at any (or all) of these periods. And that is where harmonics come into the picture.

Harmonics are familiar to many people from the physics of music. Consider the taut string of a guitar or other stringed instrument. When at rest, the string essentially traces a straight line. If plucked, it may vibrate in a very simple fashion. The centre of the string moves first in one direction (out to some limit), then back (past its position of rest), and out to some limit in the opposite direction. (The rest of the string

Photo by Tom Koppel

Photo by Tom Koppel

This series of photos shows the rising tide at the industrial wharf in Hantsport, Nova Scotia. The bulk carrier arrives at low tide, and as the tide rapidly rises, workers load the carrier with gypsum. By the time it is loaded, the carrier sits much deeper in the water, and must depart quickly before the tide falls again.

is curved accordingly as the string moves back and forth.) And it does this many times per second, which is its natural frequency. This very simple motion, with the string vibrating back and forth along its entire length, is its fundamental frequency, also called its first harmonic. The wooden body of the guitar may be designed so that it resonates with the vibrations of the string. This amplifies the sound produced by the string. Something similar occurs with the vibration of the column of air in the pipe or tube of a wind instrument like an organ. The air vibrates at a natural frequency that depends largely on the length of the particular pipe.

Strings on instruments (and columns of air in organs or woodwind instruments such as flutes or clarinets) may also vibrate at higher frequencies than the first harmonic. These higher frequencies are also called harmonics. Instead of vibrating as a whole, the string then vibrates in a number of equal parts along its length. While one part of the string is moving in one direction, the adjacent part of the string is moving the opposite way. And the periods of vibration are proportional to the lengths of these parts. These higher frequencies of vibration are always simple (whole number, or integral) multiples of the fundamental frequency, for example twice the frequency, three times, four times, etc. In music, these are the higher harmonics (sometimes called overtones). In fact, many instruments are designed to produce a series of overtones that are superimposed on the first harmonic. The body of the instrument resonates at a variety of frequencies, but they are always whole number multiples of the fundamental frequency.

What does all of this have to do with the tides? Quite a lot, because like a tide wave, harmonics are sinusoidal oscillations; that is, the motion of a guitar string follows a simple sine wave pattern. Although we cannot see air the way we can see a guitar string, the vibrations of a column of air in an organ or woodwind instrument can also be represented as a sine wave. And it is a basic feature of waves and other periodic motions (including radio waves, sound waves, and very long waves in the sea, such as the tides) that they can combine to reinforce each other if their periods happen to be synchronized — that is, if they are in phase with each other.

Picture a graph of two typical sine waves, starting at the left and with their alternating crests and troughs extending out horizontally to the right. If the timing of the crest of one wave (represented by its horizontal position) matches the timing of the crest of the second wave, the amplitudes of the two crests will be added together to form an even higher combined crest. As already mentioned, if you drop a pebble into a pool of water, you can set up a wave that radiates outward as a series of successive crests and troughs. If you drop another pebble into the water at just the right time, you can create a new wave with crests and troughs that reinforce the first series. The waves are in phase, so the crests will be higher and the troughs deeper.

On the other hand, two waves can dampen each other if their periods happen to be out of phase, thereby creating interference between the waves. In that case, the trough of one wave is "subtracted" from the crest of the other wave, resulting in a combined wave amplitude that may be less than that of either original wave. The crests will be lower and the troughs less deep. And what is true for two waves is also true for three, four, or more. Depending on their phases, the amplitude of the combined wave can be larger or smaller than any of the constituent (or component) waves. And yet, the physics of music (and other oscillating systems) shows that the various constituent waves (the harmonics) all coexist. The body of a guitar can resonate to any and all of the vibrations of the strings, not merely to some combined wave that represents their amplitudes and phases added together.

Likewise, large bodies of water on Earth can resonate in rhythm with the periods of different tide-generating forces. It is almost as though each body of water "chooses" which periodic tide-generating force(s) it prefers to respond to. (The "choice" is actually inherent in the properties of the body of water, and it is dependent on length, depth, and other factors such as friction along the shoreline or seabed.)

So, just as a pendulum swings with a natural period, bodies of water (ranging in size from a bathtub right up to an ocean basin) have natural periods of oscillation that depend mainly on the water's depth and the distance between the sides of the tub or the shorelines of the ocean. If you time it right and keep pushing at the water in a bathtub, eventually it will build up a periodic rhythm, resulting in a seesaw-like wave (or seiche) that may even slosh over the side of the tub. Or, let's go back to our analogy of the parent pushing a child on a swing. If the parent gives an added push at just the right time, the child's arc will reach higher with each push. In similar fashion, a sea or ocean basin will have a natural period of oscillation that makes it respond to the timing of particular periods (or rhythms) in the tide-generating forces.

If the natural period of a large body of water, such as an ocean basin, happens to coincide closely with the repeated gravitational impulses provided by a particular tide-generating force, the water will respond by developing a rhythmic (wave-like or oscillating) motion that is in resonance with that force. One such period might be the twelve hours and twenty-five minutes between the Moon's passage of the meridian on one side of the Earth and its passage on the opposite side as the Earth rotates. When the ocean responds primarily to the gravitational force of the Moon acting with that period, as it does in most of the North Atlantic, the result is two semi-diurnal tides in just over one day.

This period of a little over twelve hours is not, however, the only period at which the gravitational forces of the Moon and Sun can create rhythmic energetic boosts that set a body of water into motion. It was the one at the focus of Isaac Newton's tidal theory, as visualized in the tidal bulges he depicted sweeping around the world. At many places on Earth it is also, in fact, the dominant tidal period. But there is a whole array of other lunar and solar periodic cycles. Any of these other cycles can be in resonance (or close to resonance) with a particular body of water and, therefore, have a powerful influence on the tides in that sea or ocean.

Each cycle has its own distinctive period. Some of the periods approximate half a day and tend to generate semi-diurnal tides. Others are about one day long — the time it takes for the Earth to rotate once. They tend to generate diurnal tides. As we shall see, there are much longer cycles as well that affect the tides. And harmonic analysis is a system that takes all of these cycles into account to make future tide predictions. As the website of the U.S. National Oceanic and Atmospheric Administration puts it, the basic principle underlying harmonic analysis is that "any periodic motion or oscillation can be resolved into the sum of a series of simple harmonic motions."

Astronomy and Tidal Prediction

In the late eighteenth and early nineteenth century, the French genius Pierre Simon, Marquis de Laplace (sometimes called the "French

Newton") published his *Exposition du Système du Monde* ("*Account of the World System*") and later his five-volume masterwork, *Traité de Mécanique Céleste* ("*Treatise on Celestial Mechanics*"). As was the case with Newton's *Principia*, earthly tides represented only a part of Laplace's overall focus, which encompassed gravity, motion, and orbital relationships among heavenly bodies in the entire solar system. However, he found the study of tides to be a particularly interesting challenge and called them "the thorniest problem in modern astronomy."

Taking all the astronomical cycles into account, Laplace applied sophisticated mathematics to the geometry of the solar system, including the rhythmic, cyclical gravitational impulses that we call the tide-generating forces. He built on Newton's ideas but introduced much greater complexity and more subtle effects. Identifying how the tide-generating forces vary as the configuration of the Sun-Earth-Moon system changes over time (daily, fortnightly, and over longer periods), he attempted to specify how bodies of water might respond to those forces. Unlike Newton's focus on deep and unobstructed oceans, his analysis encompassed seas of limited size and depth and took into account the friction and inertia of water. Laplace, who died in 1827, was working within the conceptual framework of the progressive wave theory. He concentrated on analyzing how the rhythmic tide-generating forces could set up waves that propagated in a linear fashion along a shoreline or across an ocean.

By the mid-nineteenth century, however, progressive wave theory had shown itself to be incapable of accounting for the Atlantic tides. Robert Fitzroy and George Airy had suggested that the tides represented oscillations across entire oceans. William Whewell, too, eventually came to believe it unlikely "that the course of the tide can be rightly represented as a wave travelling [south to north] between Africa and America. We may much better represent it as a stationary undulation, of which the middle space is between Brazil and Guinea in which the tides are very small, as at St. Helena and Ascension [island]."

How to apply this concept of oscillating ocean basins to the important practical problem of making tide predictions was not at all obvious, however. Whewell's colleagues John William Lubbock and Sir Francis Beaufort (at the Admiralty's Hydrographic Office) were finding it feasible to create reasonably accurate tide tables for a small number of the most important British ports, but only after compiling records over many years for each port. This did not require any fundamental advances in tide theory, especially around Britain and France, where the tides were mainly quite regular and semi-diurnal. Local observation would reveal the establishment of any port being studied (that is, by how many hours its high tide generally preceded or followed the Moon's passage of the meridian). From astronomy, which was already an exact science, it was known how near to or far from the Earth the Moon would be at any future date, and what its declination would be. The future position of the Earth relative to the Sun, which affects the cycle of spring and neap tides, was also known. By observing how these factors affected the size of the tides over a number of years, predictions of future tides could then be made. The basic pattern of tides for any single place essentially repeats itself after 18.6 years — the length of the Moon's cycle of declination. This meant that nineteen years of observations were sufficient to create useful tide tables for key ports.

But compiling better tide tables did not lead to any more fundamental understanding of the nature of tides themselves, or their pattern on the larger scale of entire oceans. If they were not progressive waves, but oscillations set up in each ocean or other large body of water, what factors determined their range and exact timing at various places along the world's coastlines? And how could the world's navies and port authorities go about creating more accurate tide prediction tables for hundreds or thousands of places, ideally without having to spend nineteen years studying each remote or relatively minor harbour or anchorage?

This matter of tide prediction remained a major problem right into the early twentieth century. By then, acceptable tide tables existed for key ports on the coasts of Europe and North America, but the rest of the world remained largely *terra incognita* where tidal forecasts were concerned. In 1898 George Darwin delivered an alarming assessment of the inaccuracy of tidal predictions elsewhere in the world. "As things stand at present, a ship sailing to most Chinese, Pacific or Australian ports is only furnished with a statement, often subject to considerable error, that the high water will occur at so many hours after the moon's passage [of the meridian] and will rise so many feet. The average rise at springs and neaps is generally stated, but the law of the variability according to the phases of the moon is wanting," he said. "But this is not the most serious defect in the information, for it is frequently noted that the tide is 'affected by diurnal inequality,' and this note is really a warning to the navigator that he cannot foretell the time of high water within two or three hours of time, or the height within several feet [roughly one metre]."

Two or three hours? Several feet? As we saw from the critical timing of loading a gypsum bulk carrier on the Bay of Fundy, that was not nearly good enough.

Laplace Analyzes the Tides Mathematically

Putting tide prediction onto a more systematic theoretical and practical footing was a project that involved George Airy, George Darwin, and other intellectual giants of the Victorian era — the most notable of them was William Thomson (later Lord Kelvin) — and it took decades. During the second half of the nineteenth century, it became apparent that the most promising approach lay in applying the concepts of harmonics and resonant periods. It was here that Laplace had done pioneering work. He had shown that resonance was one of the keys to analyzing the motions of planets and moons in the solar system. And he had laid the conceptual groundwork for a new way of analyzing the tide-generating forces.

As we have seen, harmonics in music involves the natural frequencies of vibration of objects like guitar strings. These frequencies are related to each other by whole number ratios. The first and second harmonics, for example, have a 2:1 frequency ratio. The second and third harmonics have a 3:2 frequency ratio. And so on.

Other physical systems (for example, in the fields of electricity and quantum mechanics) are also governed by harmonics. The role of harmonics in astronomy is particularly intriguing. Laplace showed that planetary motion — the orbital properties of the planets and moons in our solar system — was by no means random. Rather, the orbital periods of the planets around the Sun, or of moons around very large planets, often display resonant and harmonic relationships. For example, Saturn's moon Enceladus makes two orbits for every one by Dione. A 2:1 resonance. The orbital periods of three large moons of Jupiter (Ganymede, Europa, and Io) are in a relationship that astronomers call Laplace resonance. The ratio of their orbital periods is 4:2:1. Similarly, Neptune and Pluto are in resonance. Neptune goes around the Sun three times for every two orbits of Pluto. And there are two recently discovered giant planets revolving around a star that is fifteen light-years away from us. The ratio of their orbital periods is 2:1. These patterns are not accidental. The orbiting bodies interact with each other gravitationally every time they swing around to the same relative position (or in some cases every second, third, or fourth time around). If they should begin to deviate from the harmonic norm, gravity will force them back into line. Their interactions serve to stabilize the orbital system.

Like the movements of distant planets and moons, the tides are also governed by the gravity of astronomical bodies that move in rhythmic, cyclical fashion. But the relationships are incredibly complicated, and they continuously change in such complex ways that they appear, at first glance, to defy mathematical analysis.

For example, the Moon's orbit is not a circle; it is an ellipse. The Moon's distance from the Earth changes continuously over the course of each month, moving from apogee (its greatest distance) to perigee (its closest pass to the Earth) and back again. That cycle repeats itself every 27.55 days. (Just how eccentric the Moon's orbit is — whether the ellipse is more or less "stretched" as compared to a perfect circle — also varies slowly over the course of a cycle lasting 8.8 years.)

Likewise, the declination of the Moon varies greatly over the course of each month. It moves from north of the equator to south of the equator and back again every 27.32 days, or roughly once a month. So it spends two weeks north of the equator (when those of us who live in the northern hemisphere see it high in the sky) and two weeks south of the equator. During the days when the Moon is farthest north or south of the equator, one of the high or low tides each day will be tend to be higher or lower than the next high or low tide. And twice a month, when the Moon crosses the equator, the two sets of tides will be nearly equal. In other words, the degree of diurnal inequality is not fixed. Like everything else in the Sun-Earth-Moon system, it changes with time. (The maximum extent of the Moon's declination to the north or south varies very slowly between 18.3 degrees and 28.6 degrees and back again over a cycle of 18.6 years. In 2005, the Moon's declination was at its minimum and is now increasing each year.)

Both of these lunar cycles (changing eccentricity and declination), along with similar cycles involving the Earth's revolution around the Sun, occur simultaneously, resulting in a pattern of mind-boggling overall complexity. Hence, when Laplace turned his attention to the tides, he realized that he had to separate out the cycles and treat each one on its own. After his death, this approach was adopted and advanced much further during the latter nineteenth century, and it came to be called harmonic analysis. (Laplace himself did not use that terminology.)

To simplify matters, in his mathematical analysis of the tides, Laplace isolated the various motions of the Sun-Earth-Moon system that generate tides and treated them as separate tidal components (also called constituents). He grouped them into species of tides, such as those with periods approximating half a day, those with periods of roughly one day, and others with much longer periods. Each component was represented by a simple sine wave with its own unique period, amplitude, and phase. These tidal components — a limited number of sine waves — could then be aggregated to create a theoretical combined wave that represented the total effect of the tide-generating forces as they act over time. Because each component constitutes only part of the larger tidal picture, they are sometimes called partial tides.

Laplace's other innovation, in part of his analysis, was to eliminate the Earth's rotation on its axis and its movement around the Sun. Instead, he imagined that the Earth stood still. In place of the Earth's eastward rotation and its orbit around the Sun, he treated the Moon and Sun as imaginary satellites that revolve westward around the Earth in the plane of the equator. For simplicity, in his mathematical system they are also highly idealized satellites, in the sense that they revolve at constant speeds and in perfectly circular orbits. Each satellite has its own period of revolution, which is its most distinctive property. He called these imaginary satellites *astres fictifs* ("fictional celestial bodies").

Laplace provided the basic intellectual framework for harmonic analysis: imaginary satellites that produce partial tides, each represented by a simple sine wave. But this was prior to the concept of resonant ocean basins. The challenge of combining the two ideas and applying them to practical tide predictions was not taken up until the 1860s by the British mathematician and physicist William Thomson (Lord Kelvin), one of the great practical technologists of modern times. A prodigy who published his first mathematics paper at sixteen and whose career was based mainly at Glasgow University, Thomson is best known for his studies in thermodynamics (low temperature scales and degrees Kelvin). He also made major contributions in electricity, magnetism, the design of undersea Atlantic cables, and improving the marine compass.

On behalf of the British Association for the Advancement of Science, Thomson established and chaired a committee in 1867 "for the purpose of promoting the extension, improvement and harmonic analysis of tidal observations." Most members were fellows of the Royal Society, but the Astronomer Royal and representatives of the Admiralty were included. Thomson outlined his basic approach as follows: "The height of the water at any place may be expressed as the sum of a certain number of simple harmonic functions of the time, of which the periods are known, being the periods of certain components of the sun's and moon's motions. Any such harmonic term will be called a tidal constituent, or sometimes, for brevity, a tide."

Thomson embraced Laplace's idea of imaginary satellites revolving around a non-rotating Earth in the plane of the equator at constant speeds and in perfectly circular orbits. This became the centrepiece of harmonic analysis.

Consider that the real Moon appears at (nearly) the same place in our sky every twenty-four hours and fifty minutes, as the real Earth rotates on its own axis. The real Sun takes twenty-four hours for one apparent circuit. To model this situation, harmonic analysis posits an imaginary Moon that revolves around the imaginary non-rotating or stationary Earth every twenty-four hours and fifty minutes and an imaginary Sun that revolves in twenty-four hours.

Harmonic analysis eliminates Newton's twin tidal bulges by making a further simplifying assumption, namely that each imaginary satellite produces its gravitational (tide-generating) effect only once, not twice, for each revolution around the stationary Earth. See Figure 10.1, which shows how two of the many possible imaginary satellites revolve around the Earth, each one exerting its tide-raising force on the ocean beneath it. (Of course, the real tide-generating forces act on both sides of the Earth simultaneously, producing in most places two tides a day.) But this additional simplification was extremely fruitful. By the late eighteenth century, wave theory had shown that water responds differently to energetic disturbances depending on the depth and dimensions of a body of water and the periods of the disturbances. A gravitational pull that reaches its peak roughly every twelve hours has a different effect from one that peaks every twenty-four hours. Having only one tide-generating impulse for each imaginary revolution allows tidal analysts to separate out the diurnal from the semi-diurnal components.

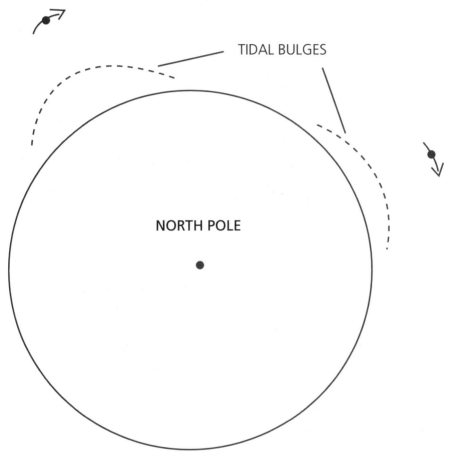

TIDAL BULGES

NORTH POLE

Figure 10.1: Satellites exerting tide-raising force.

Therefore, to take into account the twin tidal bulges and the distinction between diurnal and semi-diurnal tides, harmonic analysis creates a second imaginary satellite for the Moon. This one revolves around the stationary, non-rotating Earth twice as quickly as the first, in twelve hours and twenty-five minutes. And it creates another imaginary satellite for the Sun, one that revolves in twelve hours. In other words, there are *two* imaginary moons and *two* imaginary suns. The former ones, with periods of twenty-four hours fifty minutes and twenty-four hours respectively, generate diurnal tides. The latter ones generate semi-diurnal tides. Modern tidal science denotes the partial tides generated by these imaginary satellites as O 1 (for the diurnal lunar component), M 2 (for the semi-diurnal lunar component), P 1 (for the diurnal solar component), and S 2 (for the semi-diurnal solar component). (The numeral 1 denotes one tide a day, 2 denotes two tides. *M* stands for Moon, *S* for Sun. *O* and *P* are arbitrarily chosen notations.)

The motion of each of these imaginary satellites can then be represented as a sine wave for which the height of the wave's crest is a measure of the maximum amplitude of the satellite's tide-generating force. The period of the M 2 component is twelve hours and twenty-five minutes (the time taken by each of Newton's lunar bulges to go a little more than halfway around the Earth), while the S 2 component has a period of exactly twelve hours (because that is half of one rotation of the Earth). The maximum amplitude of the M 2 wave if it were acting on a theoretical Newtonian ocean would be a little more than twice as great as that of the S 2 wave. (This reflects the Moon's greater overall gravitational pull. The Sun's effect is 46.6 percent as strong as the Moon's.) But on the real Earth with real oceans, the actual amplitude of the wave or oscillation at any particular place will depend on the resonance of seas and oceans as they respond to the periodic energetic impulse. That resonance will vary according to such factors as the depth of the water, the dimensions of the ocean basin, friction along the seabed, and the latitude of the place in question.

Now, imaginary moons and suns revolving around the Earth in the plane of the equator do not come close to representing all of the diverse and cyclically changing gravitational configurations involved in the tide-generating forces. The real Moon, for example, has a somewhat elliptical orbit. Its distance from Earth changes, and as it does, the strength of its gravitational pull on the Earth's waters changes as well.

Put another way, as the real Earth rotates, the hypothetical tidal bulges continuously get just a bit larger or smaller each time they pass any point on Earth, depending on the changing distance to the Moon. To take this into account, harmonic analysis creates two *more* entirely separate imaginary satellites that also revolve around the Earth at the equator. One graphically represents a mathematical model of how the Moon's changing distance affects the diurnal tides, the other how it affects the semi-diurnal tides. Modern tidal notation calls them N 1 and N 2 respectively. Like the preceding imaginary satellites, they also can be represented by sine waves with distinctive periods and amplitudes. The period of the N 2 wave, for example, is twelve hours and thirty-nine minutes, and its theoretical maximum amplitude (if it were generating tides on a deep and unobstructed Newtonian ocean) is 19.2 percent of the amplitude of the M 2 wave. Why? Because that is how much the Moon's gravitational impact on the hypothetical Newtonian ocean varies over the course of a month as the Moon moves successively farther away and then closer again.

Laplace also recognized terdiurnal lunar tides, that is, components with a period of approximately eight hours (in other words, a frequency of three tides per day). And modern tidal analysis includes consideration of quarter-diurnal lunar components, with periods of about six hours (four tides per day). These are designated M 4, and they can be significant in shallow-water areas. Looking at these simple whole numbers — 1, 2, 3, 4, denoting frequencies of one, two, three or four tides a day — the analogy to harmonics in music is apparent. Depending on their dimensions, depth, and bottom friction, bodies of water can resonate at any of these frequencies.

Other imaginary satellites — all of them following idealized orbits around the equator of a stationary Earth — had to be created to represent cyclical changes in the declination of the Moon (to the north and south of the equator) and the declination of the Sun (relative to the equator, which changes with the seasons). There also had to be imaginary satellites for the changing angles between the Moon and the Sun in the sky. This is because when the Moon and Sun are pulling from the same (or nearly the same) direction, the combined gravitational effect is much greater than when they are pulling at an oblique angle to each other or from opposite directions. And this angle changes throughout each fortnightly cycle of spring tides, to neap tides and back again. There are also cycles in the tide-generating forces that extend longer than a year, such as changes in the Moon's maximum declination over an 18.6-year period. Once again, each of these satellites is represented by a sine wave with its own distinctive amplitude and period.

Table A gives a summary of the seven most important constituents. Taken together, they generally account for more than 90 percent of the total tide-generating force. Their periods are derived from astronomical observations of great accuracy.

TABLE A

Symbol	Period in Solar Hours	Description
M 2	12.42	Main lunar (semi-diurnal) component
S 2	12.00	Main solar (semi-diurnal) component
N 2	12.66	Lunar (semi-diurnal) component due to the variation in moon's distance
K 2	11.97	Soli-lunar (semi-diurnal) component due to changes in the declination of both sun and moon throughout their orbital cycle
K 1	23.93	Soli-lunar (diurnal) component due to changes in the declination of both sun and moon throughout their orbital cycle
O 1	25.82	Main lunar (diurnal) component
P 1	24.07	Main solar (diurnal) component

Combining the Many Tidal Components

It may be far from obvious how positing a large number of fictional satellites, however simply each one may be represented, furthers the task of tidal analysis and prediction. The answer constitutes a kind of good news/bad news story. First, the good news. The beauty of the system is that the sine waves representing individual imaginary satellites are simple and unchanging over time. They can be added together to create an aggregate predicted tide curve (which will not necessarily resemble a sine wave) that represents the sum of the tide-generating forces as they change over time, with each one weighted to play its proper role in the tidal prediction. In other words, once the features of the main tidal components have been determined — their period, amplitude, and phase — they can be combined to derive a picture predicting how the aggregate tide will behave, and it is possible to extend that pattern out indefinitely into the future.

The bad news is that the amplitude and phase of the components cannot be known a priori from general astronomical observation. They must be determined for every individual port of interest. This brings us back to the concept of resonance. Depending on their depth, size, and other factors (such as latitude), two ocean basins or seas may respond differently to the same tidal component. Some bodies of water may have natural periods of resonance that are closer to twenty-four hours, rather than twelve hours, in which case the diurnal components may dominate over the semi-diurnal ones. In other words, their amplitudes will be larger. It is also crucial to know the phase of each component, i.e. precisely when its amplitude is at the maximum, because this determines the outcome when two or more waves are added together. Two or more tidal components may augment each other one day and interfere with each other a few days later. (For example, for San Diego, note in Figure 10.2 how the shape of the predicted tide curve of all constituents changes from September 4 to September 5.)

In addition, the tides at any port may reflect the combined impact of tide waves coming from more than one direction and from more than one major channel, sea, or ocean basin. Hence, the amplitude and phase of each component must be obtained for every specific place, and this is teased out of the tidal data through laborious calculations after carefully

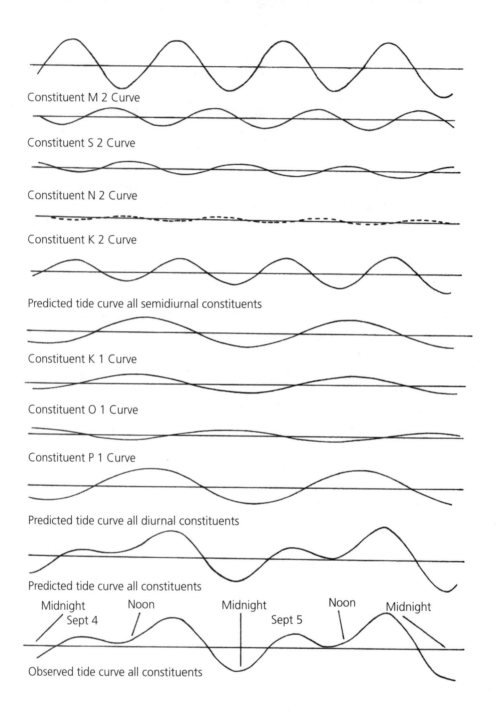

Constituent M 2 Curve

Constituent S 2 Curve

Constituent N 2 Curve

Constituent K 2 Curve

Predicted tide curve all semidiurnal constituents

Constituent K 1 Curve

Constituent O 1 Curve

Constituent P 1 Curve

Predicted tide curve all diurnal constituents

Predicted tide curve all constituents

Midnight   Noon   Midnight   Noon   Midnight
/ Sept 4   /   Sept 5   \

Observed tide curve all constituents

Figure 10.2: Major tidal constituents, with predicted and observed tide curves at San Diego, California. Courtesy of the Center for Operational Oceanographic Products and Services, U.S. National Oceanic and Atmospheric Administration.

measuring (using automated, self-registering tide gauges) the observed rise and fall of the tides and their timing at that place over an extended period. Since the period of each partial tide is different (and precisely known from astronomy), that partial tide's unique individual contribution to the total observed tide curve for the place under consideration will eventually show up in the (carefully measured and recorded) rise and fall of the total tide — but only if observations are continued for long enough. In late-nineteenth-century Britain, for the main components this was usually at least two months. Many of the lesser components require observations of one to two years, because their effects on the observed tides are so subtle.

To take a real-world example, at Hantsport on the Bay of Fundy (where my wife and I watched the bulk carrier being loaded with gypsum as the tide rose) the M_2 tide is far more dominant than simple Newtonian theory would indicate. As mentioned, the tidal bulges generated by the Moon would, for a hypothetical deep and unobstructed ocean, be a little more than twice the size of those generated by the Sun. (The amplitude of this hypothetical S_2 is 46.6 great as great as that of M_2.) However, the Bay of Fundy's length and depth give it a natural period of oscillation that happens to match very closely the M_2 period of the tides in the open Atlantic. This means that roughly every twelve and twenty-five minutes, the tide in the Bay of Fundy gets an extra energetic boost from the Atlantic tides, making it surge in with great force.

In addition, the bay has a somewhat funnel-like shape, becoming narrower and also much shallower toward its closed northeastern end, which further augments the height of the tide at that end, just as waves build into high breakers as they move into shallow water when they approach a beach. So the flood tide surges in and then recedes, only to surge in again in almost perfect resonance with the next flood tide in the open Atlantic. This is why the Bay of Fundy has often been called the world's largest bathtub. The situation is similar on the Bristol Channel (Severn Estuary) in Britain and the Gulf of Saint-Malo in France, where the abbey of Mont Saint-Michel is located.

The link between the periods of the partial tides and their amplitudes in the Bay of Fundy is dramatic. At Hantsport, the M_2 component produces a mean tidal range of 5.665 metres (almost 19 feet). It would be even more on a particularly large spring tide. The S_2 component (which, according to Newtonian theory, should have 46.6 percent of the amplitude of the M2 component) raises the tide at Hantsport only an additional 0.847 metres (less than 3 feet). And the impact of the diurnal tides is even weaker. The K 1 component, which elsewhere in the world is usually the strongest of the diurnal partial tides (and in Newtonian theory would be 58.4 percent as strong as M 2), raises the tide at Hantsport by only 0.209 metres (about 8 inches). When these components are added together, we can see why the tides on the Bay of Fundy are semi-diurnal (with only a very small diurnal inequality) and highly regular. Each high and low tide is nearly the same as the previous or succeeding one. They vary only gradually, day by day, as the M 2 tide in the open Atlantic goes through the spring to neap cycle. Tides on much of the coasts of Britain and France follow the same overall pattern. Figure 10.3 represents the rise and fall of the tides over a one-week period at Plymouth on Britain's southern coast in June 2003. There are two tides a day, and the difference between successive ones is not large, although the fortnightly spring to neap cycle is apparent. Notice also that the time of each high and low tide comes a bit later — by about fifty minutes — each day over the course of the week. A graph of the tides on the Bay of Fundy over the course of a week would look almost identical.

By contrast, consider Victoria, on British Columbia's inner south coast, which has a mesotidal range of tides, as shown in Figure 10.4, the tidal calendar for August 2005 at Victoria. (Maximum tidal range at spring tides is about 3.3 metres, or under 11 feet, and the mean tidal range is much less.) The largest component at Victoria is K 1, the diurnal partial tide resulting from the changing angles of declination of the Moon and Sun. It accounts for a mean tidal range of 0.627 metres (about two feet), while the M 2 component accounts for only 0.373 metres (just over one foot). The next largest contribution to the tidal range at Victoria is made by O 1, the lunar diurnal component, which has about the same strength as M 2, adding 0.370 metres (1.2 feet) to the mean tide. S 2, the solar semi-diurnal component, only contributes an additional 0.102 metres (4 inches). When these components are added together, they produce a mixed tide in which the diurnal components are dominant. The sea responds mainly to the daily gravitational impulses, rather than to the twice-daily ones. Notice how different this looks from the pattern of tides at Plymouth, England. There are often two tides a day, but for much

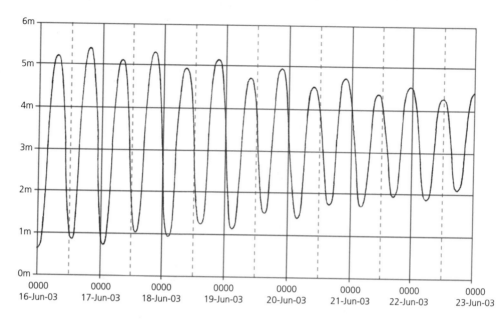

Figure 10.3: *Tides at Plymouth, England, from midnight local time (+0000 GMT) on June 16, 2003, to June 23, 2003. Times do not take Daylight Saving Time into account. © Crown Copyright and/or database rights. Reproduced by permission of the Controller of Her Majesty's Stationery Office and the UK Hydrographic Office.*

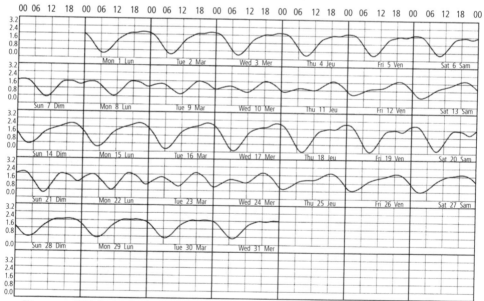

Figure 10.4: *Tidal calendar for Victoria, British Columbia, for August 2005. CHS tidal information reproduced with the permission of the Canadian Hydrographic Service.*

of each month one of the highs or lows is so much greater than the other that the difference between a high tide and a low one is barely discernible. Sometimes the tide curve nearly levels out, remaining high (or low) for eight to ten hours. An observer would hardly notice any change at all, so on many days there is really only a single ebb and a single flood.

Along most of the rest of the British Columbia coast, which extends northward from Victoria, although the tides are also mixed, the semi-diurnal tides predominate. This is certainly true in B.C.'s Gulf Islands, where I live, even though we are only about forty kilometres (twenty-five miles) north of Victoria. On most days of each month there are two clearly distinguishable high tides and two lows, even though those highs and lows are not nearly of equal amplitude. And this pattern of mixed, but predominantly semi-diurnal tides holds true for most of the Pacific coast of North America, from Seattle and Vancouver down to southern California. The tides around Victoria are an anomaly that underscores just how greatly the pattern of tides can vary between places only a short distance apart.

West Coast tidal patterns also demonstrate how different the *phase* of the tides can be at places only a short distance apart as the crow flies. Dr. Louis Druehl, the leading British Columbia expert on kelp and other seaweeds, lives at Bamfield on the west coast of Vancouver Island, facing the open Pacific. He has found that he can collect samples of kelp at low tide near Bamfield and then make the several-hour-long drive across Vancouver Island to collect kelp at low tide on B.C.'s inner coast. The phase (or timing) of the tides on the two opposite shores of Vancouver Island is different by as much as five hours, because it takes that long for the tide wave to propagate from the open Pacific into the constricted (and in many places shallow) passages of the inner coast.

An even more extreme situation is found in the microtidal diurnal tides on the Gulf of Mexico. Figure 10.5 shows a week of tides at Biloxi, Mississippi. The sea there responds almost entirely to the diurnal tidal components, and hardly at all to the semi-diurnal ones. There is only one high water and one low water each day, although a hint of semi-diurnal influences can be seen in the slight "saddle" shape at high tide on June 21.

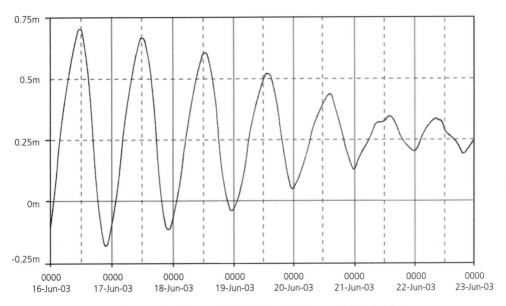

Figure 10.5: One week of tides at Biloxi, Mississippi, from midnight local time (+0600 Greenwich Mean Time) on June 16, 2003. © Office of Coast Survey, US National Oceanic and Atmospheric Administration.

William Thomson's Clever Invention

William Thomson realized that harmonic analysis could only be applied in practice by determining the amplitude of each tidal component, as well as its phase, for every port of interest. Once these were known, the tides for that place could be predicted well out into the future. Anywhere from seven or eight to twenty or more components might be required. (Predictions in shallow seas and estuaries required more components.) But in the days before digital computers, the hand calculations involved were prohibitively tedious.

Thomson, who was a brilliant inventor with many patents to his name, came up with a solution. Taking up a suggestion by his brother, James Thomson, and turning it into a practical device, he designed a novel tide-predicting machine in 1872. It was a cumbersome, Rube Goldberg–style contraption with many interconnected pulleys that were free to slide up and down along vertical tracks, one pulley for each tidal component. Each pulley rotated at a speed that was scaled according to the period of the component it represented. The phases and amplitudes for a particular port were set in advance of running the machine by adjusting the starting point (relative height) of each pulley on its track. Then, by means of a wire that passed over each pulley in turn, as the pulleys rotated and slid up and down they guided a stylus that automatically traced out a curve on a rotating drum; this physically summed the sine waves of the tidal harmonics. The first device built, an early mechanical computer, could sum ten harmonic components, giving each one its proper weight for the specific location.

Once the amplitude and phase for each component had been determined for a particular place, turning a hand crank would produce a series of predictions of future tides. Thomson's invention was a major breakthrough. More sophisticated versions were eventually put to use by the world's navies in the ensuing decades, although even at the end of the nineteenth century Britain still had only one really impressive functioning calculator, a device built for the government of British India. It took a couple of days of machine time to compute the high and low tides and times for a single port for each day in a one-year period. A more advanced German machine designed in the 1930s could calculate tides on the basis of sixty-two harmonic components, some of them representing long tidal cycles, such as the Moon's fortnightly and monthly motions, others representing very short-period harmonics, such as the terdiurnal and quarter-diurnal ones. Today, computers do the same job much faster and better.

Large Tides Require More Accurate Forecasts

By the late nineteenth century, harmonic analysis had brought a major advance in tidal science. Just how and why the tides behaved the way they did remained something of a mystery, however, especially as regards tides in the larger ocean basins. As observations and tidal predictions were made in more places, anomalies were discovered. These showed that some fundamental insight was still lacking.

At the very least, however, the accuracy of tidal predictions was becoming adequate for most navigational purposes. In places like the Bay of Fundy, the predictions would *have* to be much better than the rough "two or three hour" estimates that George Darwin bemoaned if they were to be of any use at all. With a tidal range of fifteen metres (some fifty feet), the tide can rise or fall so fast that mariners must know the timing very closely to avoid getting into trouble.

My wife and I saw this clearly when we went to enjoy dinner at Hall's Harbour Lobster Pound, a restaurant overlooking a rocky nook full of local fishing boats. When we arrived, the tide was high but just beginning to fall. We strolled around the scenic little harbour and took some pictures. Then we went through the procedure of selecting our lobsters from a live tank, taking them in a plastic bin to the cook shed, and waiting in the restaurant for the boiled lobsters to be served. We tied on our bibs and tucked into our feast, enjoying fries and coleslaw on the side, all washed down by good Nova Scotia microbrewery beer.

We dawdled a bit after dinner, chatting with the owner and buying postcards at the gift shop. As we left, we were taken aback by what we saw. Where the boats had been floating comfortably a couple of hours earlier, with their decks at about street level, there was now only mud. The boat decks were three or four metres below street level and their hulls were resting on the sea floor. The waters of the Bay of Fundy had vanished entirely. The tide had ebbed that fast.

Photo by Tom Koppel.

Fishing boats at low tide in Hall's Harbour, Nova Scotia.

Holding Back the Tide

How Civilizations Have Coped with High Water and Low

Geologist David Mossman pointed down at a tightly spaced row of half-rotten wooden posts that protruded from the edge of the mud flat, just below where we stood. The posts formed what looked like a low picket fence that meandered off, running along the seaward side of a roadway that was raised about two metres (some six to seven feet) above the mud flat fringing the Bay of Fundy. Mossman explained that the road, and the berm on which it is built, is actually the modern dike that keeps the bay's high spring tides from flooding a broad swath of farmland that lies behind it. The decaying wooden posts are all that remains of a much earlier dike that was built by hand a century or more ago.

We were looking out at the Cumberland Basin, just below a historic site called Fort Beauséjour in New Brunswick, the same place where Mossman had nearly perished some years earlier when caught out on the mud flats by a rising tide. Nearby, pieces of heavy machinery were doing repairs and maintenance work on the modern dike, a never-ending task. The low-lying farmland, where cattle grazed, would be ruined if high tides were able to top the dikes and flood the land with salt water. Across that land runs the main railway line between Halifax and Moncton, which also needs protection. Holding back the tide is an essential feature of life on the Bay of Fundy. As the old wooden posts show, it has been that way for centuries.

The first European inhabitants in the area were French farmers who settled today's Annapolis Valley, Nova Scotia, in the early 1600s and called their homeland Acadie. They recognized the potential richness of the tidal flats and salt marshes along the bay and all of its estuaries, where high water had deposited thick layers of organic silt for thousands of years. In an arduous decades-long project (applying the same techniques used for reclaiming tidal marshland in Europe), they built dikes to prevent the highest spring tides from flooding those areas.

As we could see, the front line of the old dikes, facing the sea, often consisted of wooden posts driven into the mud flat or salt marsh. In behind the posts the Acadians dumped rocks, wood, and dirt to create a raised line of fill, and they planted sod on top. Elsewhere, the dikes were mainly earthen structures about one and a half metres (five feet) high, with veneers of salt grass sod on both sides to help hold the soil in place. Spaced every few hundred metres, the Acadians installed discharge sluices (called *aboiteaux*), which were large wooden pipes that ran under the dike. On the seaward end of each pipe, there was a simple wooden "door" on leather hinges. It was a type of flap valve that allowed fresh water to drain out from the landward side after heavy rains, but did not let salt water intrude from the seaward direction. Winter rains and melting snow flushed the salt out of the diked-off areas within a few years, creating thousands of hectares of unusually fertile land, where the Acadians prospered.

But Britain and France were rival colonial powers in the region. The original Fort Beauséjour was built by the French in 1751. In 1755, with the two countries at war, the British expelled the Acadians. Some were shipped back to Europe. Others, their name shortened to "Cajuns," ended up in Louisiana. Some hid near the borders of Quebec and emerged to resettle in northern New Brunswick, where their descendants still live today. British settlers, including many Loyalists who moved north from New England during the American Revolution, took over the rich

Acadian lands and maintained the existing dikes. Dike repair and maintenance continues today and will for the foreseeable future.

Dikes and Seawalls

Building dikes along the uppermost portion of the intertidal zone is just one of the many ways that coastal engineering has been applied to cope with high tides or to exploit the rise and fall of the tides. Other ways include placing fixed tidal barrages (a type of dam), or more sophisticated (and costly) tide and storm surge barriers that can be opened and closed, across the mouths of estuaries. Both of the latter approaches are intended to keep particularly high tides, tidal bores, or storm surges from sweeping up the rivers and flooding upstream farmlands. On five rivers flowing into the Bay of Fundy, barrages were built between the mid-1950s and the early 1970s. In part, they serve as substitutes for the more costly maintenance of endless kilometres of riverside dikes. These Canadian projects were undertaken after regional officials visited Holland to see how the Dutch were dealing with dangerous episodes of high water. And similar concepts have been applied in many other places.

The earliest known dikes were built by the agrarian Harappan people in the Indus Valley more than two thousand years before the birth of Christ. (This was the same Indus Valley civilization that devised the sophisticated tidal dockyard at Lothal while the Egyptians were building the pyramids.) Dike construction and maintenance have been practised ever since by civilizations in many parts of the world.

The Dutch are, of course, particularly noted for their dikes. Large areas of the Netherlands are habitable only because of the vast dike system they have developed over the centuries. Until recently, the world's longest dike was the Afsluitdijk, which is actually a long, low dam that was built in 1932 across the entrance to the former Zuidersee, thereby creating a huge but shallow freshwater lake called the Ijsselmeer. Today, it has been trumped by the thirty-three-kilometre (twenty-mile) long Saemangeum Seawall, completed in April 2006 on the southwestern coast of Korea, an area with very large tides. This dike was controversial and opposed by environmentalists because its construction blocked off two river estuaries and a large area of natural coastal wetlands while creating 400 square kilometres (over 150 square miles) of new farmland and a freshwater reservoir.

The Dutch situation is quite different from the Bay of Fundy, where the main challenge (solved in part by dikes) is coping with the extremely high water that comes with every spring tide. On the coast of Holland, by contrast, the tidal range is not particularly large, averaging in most places well under two metres (about five feet). Much of the land, however, lies below even ordinary mean sea level. (*Nederland*, the Dutch for Netherlands, means "low land.") This is mainly because the land has been steadily subsiding as part of a long-term response to the retreat of glaciers since the end of the Ice Age. (Some parts of Northern Europe are slowly being uplifted, others are subsiding.) The gradually rising worldwide sea level, in part attributable to global warming, has aggravated the problem for the Dutch, so they have centuries of experience holding back the sea, primarily by building and maintaining dikes.

And just as dikes on the Bay of Fundy allowed the Acadians to create fertile farmland from barely productive salt marshes, in Holland huge areas that are below sea level have been artificially reclaimed by building dikes around them and continually pumping out any salt water that intrudes. These are called *polders*. One such area, the Noordoostpolder, encompasses nearly 600 square kilometres (230 square miles), all of it land that would be beneath the sea, at least much of the time, as the tide rises and falls. The dikes do a good job of holding back the sea, at least under normal circumstances. But now and then, conditions become extreme, because of either the weather or long-term tidal cycles, or both. And then coastal flooding can be devastating.

Storm Surges

When a tide table forecasts the height of a high spring tide, that prediction is only valid for times of average atmospheric pressure, and it ignores the strength and direction of the wind. If a major storm brings with it very low air pressure, the barometer falls and the waters of a high tide rise to abnormal heights. (The gravitational tide-raising forces are the same as usual, but there is less resistance to them from the atmosphere, less air pressure to "hold down" the water.) The result is a storm surge, a tide that can be several metres higher than normal. In addition, powerful storms bring with them very strong winds. If these winds blow in the direction of shore, they can drive the water shorewards, creating a natural gradient in the water that lasts as long as the winds continue at strength. This raises the sea level on the shore, an effect that is called storm set-up. Finally, such winds can also generate unusually large waves, which makes the water rise even higher when they sweep onto an open or exposed coast. This is especially true if there is a shallow and very gradually shelving profile, as with much of the Dutch shore.

During the night of January 31 to February 1, 1953, a powerful storm on the North Sea coincided with a very high spring tide. The storm surge created high water that was nearly 3.5 metres (11.5 feet) above ordinary spring high tides, and huge waves as well. The sea surged right over the dikes on the Dutch coast. Many dikes and seawalls simply collapsed, allowing extensive flooding and killing more than eighteen hundred people. An estimated ten thousand farm animals were drowned, largely on the polders, and forty-five hundred buildings were destroyed. Overnight, 1,750 square kilometres (675 square miles) of the Netherlands were covered in water.

On the coast of Britain, more than three hundred people were killed as waves swept over sea walls. Thirty thousand people had to flee their homes to escape the flooding, and thousands of homes were destroyed or severely damaged. Much of the flooding was along the lower Thames, although London itself got off with relatively modest damage, considering that riverside streets like the Embankment are barely a metre (three feet) above normal spring high tide. (An earlier combined storm surge and high tide *did* flood London in 1928, killing twenty-eight people.)

There was severe flooding and damage on the coasts of Belgium, France, and Denmark as well. Many fishing boats and ships at sea were also overwhelmed and sunk, notably the MV *Princess Victoria*, claiming more than 350 additional lives. It was a disaster, but the kind of natural event that can be expected perhaps once each century.

In both Britain and the Netherlands, the North Sea storm led to decades of planning and construction, resulting in a series of enormously expensive engineering projects that have strengthened and raised the heights of the dikes and seawalls. Today, much of the Dutch coast is protected by steel walls twelve metres (forty feet) high. But, as with the barrages across the mouths of estuaries on the Bay of Fundy's rivers, in many places it was more efficient to prevent the storm surges from reaching long and vulnerable stretches of shoreline, rather than raising the dikes to enormous heights everywhere.

The British built the Thames Flood Barrier across a 523-metre (1,716-foot) wide stretch of the river at Woolwich Reach, below London. Completed in 1982 at a cost of around US$933 million, it consists of massive hydraulically operated steel flood gates, shaped like half-cylinders, that rotate on vertical axes. During normal tide and weather conditions, they are kept open to allow ships to pass between them. When a flood threatens, they rotate to seal off the Thames and protect areas upriver from the barrier, especially London and its port facilities. Since completion, the barrier has been closed more than ninety times.

The Netherlands responded to the North Sea floods with even larger and more expensive measures. The most ambitious one is the 9-kilometre (5.6-mile) long Oosterscheldekering (meaning "eastern Scheldt surge barrier"), which spans the gap between the islands of Schouwen-Duiveland and Noord-Beveland. It effectively prevents extremely high tides from reaching the Oosterschelde estuary, which was the main mouth of the Scheldt River in Roman times and is now a large inland sea. Constructed with colossal concrete pillars up to 38.7 metres (127 feet) high and completed in 1986 at a cost of around US$3 billion, it is among the engineering wonders of the modern world. It looks a lot like a fixed dam. However, it has huge sluice gates that can be left open in normal times but closed when a flood threatens, which has happened twenty-three times. The open gates permit the passage in both directions of normal tides and marine life, which minimizes the impact of the giant barrier on the environment. (Some tidal barrages on rivers draining into the Bay of Fundy have drastically changed the ecosystems of the estuaries that are protected by them.)

The Saros Cycle of Extreme Tides

As the North Sea storm of 1953 showed, even an elaborate system of dikes and seawalls (which the British and Dutch already had at the time) can be overwhelmed if a storm surge occurs when the tide is unusually high. And exceptionally high tides are natural, recurring and predictable events with an astronomical cycle of their own. As we have seen, when its elliptical orbit brings the Moon closest to the Earth (perigee), its gravitational influence is at its strongest. When the Moon and Sun are in or very close to conjunction (times of new moon or full moon), their combined gravitational influence generates maximum spring tides. Similarly, the changing declination of the Moon and Sun produces a cycle of relatively stronger and weaker tide-generating forces. Most of the time, the peaks of these three cycles do not coincide, but every 18.03 years they do. This time of gravitational coincidence is called Saros, or the Saros cycle, from an ancient Babylonian name that has been adopted by astronomers. (Certain solar and lunar eclipses repeat themselves in accordance with the same cycle, and Chaldean priests were aware of this in 800 B.C.) At times of Saros, it is predictable that the high spring tides will be at their greatest.

If a storm surge (and the wind set-up that often accompanies it) happens to occur during a high spring tide at a time of Saros, high water can reach particularly disastrous heights. On the Bay of Fundy, this happened during the night of October 4–5, 1869. Almost a year earlier, Lieutenant Stephen Saxby, an engineer working for Britain's Royal Navy, alerted newspaper readers that the Saros conditions would prevail on October 5, 1869. At that time, "the New Moon will be on the Earth's equator when in perigee, and nothing more threatening can ... occur without miracle," he said. "In the meantime," he added, "there will be time for repair of unsafe sea walls, and for the circulation of this notice throughout the world." But the warning was largely ignored.

On October 4 a storm hit the northeastern coast of America, creating what is remembered as the Saxby Gale or Saxby Tide. Blowing from the south, by nightfall the winds reached hurricane force, and the tide kept rising. Flooding and extensive wind damage occurred in New England and Canada's Maritime provinces. Between Washington, D.C. and the upper Bay of Fundy, more than 150 vessels sank or were blown ashore. By evening, waves were breaking over wharves even within the shelter of the harbour at Saint John, New Brunswick. For people who lived without weather forecasts, phones, radios, or electric lights, it was a nightmare. As one man described it, "The extreme darkness, the constant roar and tumult of the wind, the lashing rain, the groaning of great trees, the hail of debris, shingles, branches, objects large and small, falling everywhere, roofs carried aloft, whole buildings collapsing, all gave a paralyzing sense of insecurity and calamity."

At the head of the bay, the highest tide occurred after midnight on October 5, rising about one metre (three feet) or more above the level of the dikes and flooding over them as if they did not exist. On the Cumberland Basin (where we viewed the ancient wooden dike posts) two fishing schooners were swept right over the dikes and carried five kilometres (three miles) into the salt marshes. At Moncton, New Brunswick, which is far up the Petitcodiac River from the bay, the tide peaked at two metres (six and a half feet) higher than any previous high tide on record, rising nearly three metres (nine feet) above one of the city's riverside wharves. A plaque still indicates the height of the tide that grim night.

Many people drowned, and countless farm animals and buildings were swept away. The total number of dead is not certain, although estimates exceed one hundred for Canada's Maritime provinces alone. Eleven died in the sinking of a single barquentine, the *Genii*, which was en route to Saint John. In 1979, historian Norman Creighton reported on CBC radio that "in the churchyard at Hillsborough [New Brunswick] is a whole section of tombstones raised to the victims of the Saxby 'tide'.… Farmers had gone down to the marshes, in an attempt to lead their livestock to safety, and then the dikes broke, and they were swept away by a great tidal wave."

What is the likelihood that a similar calamity will occur again? Within the next century or two it is almost inevitable. In their study of the Fundy tides, Con Desplanque and David Mossman point out that "the *Saros* cycles are long term harmonic motions. This means that near the top or bottom of the cycle the rate of change with time is relatively small. Thus, the 'peaks' of *Saros* cycles are not confined to points in time, but to rather short intervals of time." If a large storm occurs at a time of high spring tide that is even *close* to the peak of the Saros cycle, the storm surge and related effects can be devastating. Such an event was the Groundhog Day storm of February 2, 1976, which produced a huge storm surge on top of the astronomical high tide. Much of Bangor, Maine, was flooded, and water level on the Penobscot River reached 3.2 metres (10.5 feet) above the predicted tide level. At Saint John, on the Bay of Fundy, water reached 1.5 metres (4.9 feet) higher than predicted and left the waterfront in a shambles. If the storm had occurred at high spring tide a month or two later (at the peak of the Saros cycle) the effects would have been drastically worse. The next time of Saros will be in 2012 to 2013.

Aboriginal Clam Gardens on the Northwest Coast

Not all the ways that coastal people have dealt with or exploited the tides involve high tides. Low tides and the lower part of the intertidal zone can be useful as well, especially for traditional shellfish harvesting. And in recent years a remarkable example has been discovered of how aboriginal people used to "engineer" the lower intertidal zone to enhance shellfish production.

In 1995, geologist John Harper was flying in a helicopter over Baker Island, one of many in the Broughton Archipelago on the sheltered inner coast of British Columbia near the northern end of Vancouver Island. He was conducting an aerial survey of coastal habitat, focusing on the bio-physical characteristics of the intertidal zone. The knowledge gained might prove crucial in case of an oil spill. The low tide at the time was one of the lowest of that year.

Harper spotted a rock wall that ran across the mouth of a small, shallow bay. It was just below the surface of the water. As the chopper moved on to another little cove, he saw another very similar rock wall, and then another. In the course of his aerial survey, he soon noticed dozens of the unexpected stone walls. Harper was familiar with traditional Native stone fish traps, and these walls did not look at all like them. As he later told Vancouver journalist Stephen Hume, "If you saw just one, you might not even notice, but these were all over the place. The light clicked on. I said, 'Stop the helicopter.'" Upon landing, Harper found that the stones curved around the mouth of one cove in an almost perfect arc. Trapped behind the wall was a nearly level terrace of mixed sand and broken bits of shell, all full of clams and other bivalve mollusks. He considered that the walls might have been left by glacial action at the end of the Ice Age, or that they might have been formed by more recent ice action, as with the boulder barricades on the coast of Labrador. But neither hypothesis made sense. "I was pretty sure they were man-made," he said, and that "whole beaches were actually engineered." Harper launched a seven-year research project to study the phenomenon.

What he learned was that before the arrival of European fur traders and settlers, and before their populations were decimated by introduced diseases, the aboriginal Kwagiulth people in the area had built these hundreds of low retaining walls. The purpose, if it was indeed done consciously (which Harper thinks likely), was to enhance the productivity of their local shellfish beds. So they represent a form of "large-scale, pre-contact mariculture." Improved productivity was certainly the effect, as he discovered by collaborating in his study with leading Kwagiulth elders and coastal archaeologists, anthropologists, and biologists.

"Large-scale" is hardly an exaggeration. Harper mapped more than 350 of these places, which he called clam gardens or clam terraces. Some were just tiny nooks between rocky headlands, little coves full of clams and other shellfish, with stone walls as short as ten metres (thirty-three feet) long running across them at the extreme low tide line. Others were over one kilometre (half a mile) in length. Most of the walls were from one-half metre to one metre (one and a half to three feet) high. And he learned about the biology of clams in the intertidal zone. "Clams are all found in the one- to three-metre [3.2 to 9.8 foot] intertidal zone," he said — in other words, that height above the extreme low tide line. "That's the prime area for clam productivity — and all these ridges are at the precise height to optimize clam production. I've also been learning about clam culture. It appears that harvesting clams improves productivity, that constant tilling and removing large clams improves the habitat for the clams that remain."

But the process may not have been entirely conscious and intentional. Digging clams with a sharpened stick — as was traditionally done — would inevitably involve turning over any rocks that were encountered, and possibly removing them. This would make sense, because removing sizeable rocks from the clam bed leaves room for more clams. And where would a Native clam digger put the rocks? It would always be easier to carry them downhill and add them to the line of rocks at the lower edge of the intertidal zone than to carry them uphill to the beach. Over decades and centuries, systematic seasonal clam digging, which was an important part of aboriginal food gathering, would result in clam terraces with higher stone walls, deeper beds of sand and shell held back behind them, and intertidal zones with much gentler slopes than the natural gradient.

The study carried out comparisons between productivity in ordinary beaches, full of rocks and with steeper slopes, and clam gardens behind stone walls. It found that building a ridge of stone and letting the sand and shell "hash" fill in behind it increased the clam productivity by three and a half to four times. In the course of interviewing elders, a few nearly forgotten old stories and songs about working the clam gardens turned up. Clam harvesting had never been a high-prestige activity. Although it was done on a massive scale — there are endless kilometres of shell middens left behind by millennia of harvesting and processing behind every beach — it was an aspect of northwest Native culture that tended to be lost. But it appears that similar clam gardens or terraces existed on other parts of the northwest coast, including on Puget Sound in Washington State. The search has just begun for more traces of this ancient method of engineering the intertidal zone.

The Invasions of Normandy and Inchon

Seldom has the timing of the Moon and the tides been important to the fate of so many people as on the morning of June 6, 1944 — D-Day — when the Allied armies stormed ashore at Normandy to challenge the entrenched Germans. Military strategists and engineers on both sides had, of necessity, factored the tides into all of their planning.

The Germans, expecting an eventual Allied invasion, had spent several years and enormous effort reinforcing their defences along the cliffs and beaches of northern France. Artillery and machine gun emplacements, shielded by concrete, were installed all along the coast, and especially in places commanding the likely beaches where the Allies might land. The tidal range on most of the French coast of the English Channel is very large, and the slope of the mainly pebble intertidal zone tends to be gentle, making for extremely wide beaches at low tide. To make it more difficult to land on those beaches, the Germans planted minefields, constructed steel tank traps, and strung barbed wire and other obstacles far out from the seawalls and close to the low tide line.

The tides and obstacles were among the factors the Allied Supreme Commander, General Dwight Eisenhower, had to consider in planning the invasion. He and his staff decided that a successful series of landings required several conditions. One was good enough weather that the seas would not be too rough and the visibility would be adequate for bombardment by planes and ships' guns. There had to be a bright (full or nearly full) moon for large-scale paratroop and glider drops behind the coast on the night preceding the actual landings. (This advance guard would capture bridges, cut railway lines, and prevent the Germans from bringing up reinforcements.) And there had to be a very early morning low tide (it came at 5:25 A.M. on June 6), so that the rising tide would be flooding in rapidly by H-hour, the time for the first landings to begin. (The exact timing of H-hour varied, beginning as early as 6:30 at Utah Beach, an American zone, and coming as late as 7:45 on Juno Beach, where the Canadians landed. In each case, it depended on the tide conditions and beach configurations.)

A day with an early morning low (but rising) tide was considered ideal, because the low-tide obstacles placed on the beach by the Germans would still be visible. This would allow demolition parties to clear them and the first wave of landing craft to avoid them. Then, as landing craft ran aground in the shallows to offload, the rapidly rising tide would help the craft to float off again quickly, so they could get away from the beaches and back to the transport ships to pick up later loads of men and vehicles. A very early low tide in May or June also meant that there would be a second low and then rising tide during daylight hours toward the end of that first and "Longest Day," allowing for the maximum offloading of men and materiel. These requisite conditions (full moon and early low tide) only existed for about three days each month that spring. June 5 was the first date selected for D-Day, but poor weather forced Eisenhower to postpone the invasion for one day.

Another important part of the invasion plan involved getting massive amounts of vehicles and supplies, as well as additional troops, ashore in the days and weeks immediately following the initial landings, when the forces in Normandy would be vulnerable to German counterattack. Until the Allies captured a major harbour, such as Cherbourg, they would have to deliver those heavy supplies onto beaches with a large tidal range and exposed to the often furious winds and waves of the English

Channel. The initial landings involved some specially designed amphibious vehicles and tanks that could land in surf and on beaches, but not all equipment was suited to those conditions. Landing craft could be beached near high tide and unloaded when the tide dropped, but then they were stuck until the next rising tide. Allied planners did not think the follow-up effort to reinforce the beachhead could be done entirely by landing craft. A novel concept was required to get around this problem.

It was Winston Churchill himself who realized that artificial ports would have to be towed across the Channel from England and installed on the Normandy shore, an idea that led to the design of two mulberry harbours, so named because mulberry trees grow very rapidly. About ten days after the invasion, Mulberry A was operational at Omaha Beach in the American sector and Mulberry B at Gold Beach in the British and Canadian sector. Each was designed to take into account the broad intertidal zone and very large tides. As Churchill had noted in a memo in May 1943: "Piers for use on beaches: They must float up and down with the tide."

The mulberries consisted of several components, all of which had to be pre-built in secrecy in Britain — the workers fabricating them did not know what they were making — and towed across once the early landings were underway. For protection against wind and waves, the perimeter of each mulberry consisted of seventy-five prefabricated hollow concrete shells, or caissons, each about sixty metres (two hundred feet) long. When these were set end to end, filled with water, and sunk, they created long breakwaters that stood nine metres (thirty feet) above sea level at low tide and were still about three metres (ten feet) above at high tide. Leading out from the beach at each mulberry were some ten kilometres (six miles) of flexible steel roadways that floated on massive pontoons. The roads terminated at pier structures that could be jacked up and down with the tides, on huge legs resting on the seafloor. Seven Liberty ships at a time could dock at the piers within each mulberry, and seven thousand tons of vehicles and supplies could be offloaded and brought to shore each day.

The concept worked well, but a tremendous storm hit two weeks after D-Day, destroying the American mulberry. As one retrospective put it, "The Americans had to return to the old way of doing things: bringing landing ships in to shore, grounding them, off-loading the ships, and then refloating them on the next high tide." The British mulberry served well for ten months, however. Two and a half million men, half a million vehicles, and four million tons of supplies were efficiently offloaded there, avoiding all the delays entailed by dealing with the rise and fall of the tides.

A military operation that required even more precise coordination with the tides than D-Day was the American assault at Inchon during the Korean War. "The amphibious landings of 15 September 1950 were [General Douglas] MacArthur's masterstroke," wrote British journalist and military historian Max Hastings. It put a powerful U.S. force ashore far behind the lines of the hitherto victorious North Korean army, which had the South Koreans and UN allies pinned down at Pusan on the southern tip of the Korean peninsula.

The problem for MacArthur at Inchon, a port city of 250,000 with a sheltered harbour, was the 9.8-metre (32-foot) tidal range on Korea's west coast. Only on a day with an extremely high tide would the large landing craft have a window of three hours to come in and unload men against the city's seawall. At all other times, most of the harbour would be an impassable morass of mud, and it was accessible only by a single narrow dredged channel beset by treacherous tidal currents that ran at up to eight knots. Aggravating the difficulty was an offshore island, Wolmi-do, with a North Korean garrison that commanded the approaches to Inchon. The island would have to be captured first during a separate assault and landing at high tide eleven hours before the main attack on Inchon itself. This would deprive the Americans of any kind of tactical surprise for the later assault. Finally, Hastings noted, "the tide times dictated that the main landings must take place at evening, leaving the assault force just two hours of daylight in which to gain a secure perimeter ashore." In other words, there had to be a morning high tide landing at Wolmi-do to take out the garrison, followed by a long daytime wait during low tide before the main force of ships could race up the approach channel and send in landing craft on the evening high tide.

Most of MacArthur's colleagues thought he was crazy, that the timing and huge tides made the operation too tricky by half. At the crucial meeting in MacArthur's Tokyo headquarters, the Navy commanders outlined the proposed plan, but also pointed out the great difficulties entailed by the landings. "The best I can say is that Inchon is not impossible," opined Rear-Admiral James H. Doyle after delivering his briefing. But MacArthur, whose forces had hopscotched from one island to another across the western Pacific during the Second World War, had more confidence in the Navy's abilities than the Navy did itself. "Admiral," he replied, "in all my years of military service, that is the finest briefing I have ever received ... You have taught me all I had ever dreamed of knowing about tides.... The very arguments you have made as to the impracticalities involved will tend to ensure for me the element of surprise. For the enemy commander will reason that no one would be so brash as to make such an attempt.... We shall land at Inchon, and I shall crush them."

And so MacArthur did. At dawn on September 15, following heavy naval bombardment of the island, the Marines swarmed ashore at Wolmi-do and captured it with ease within an hour. Then "as the tide swept back to reveal miles of dull, flat mud between the invasion fleet and the shore, the men in the ships waited, impotent, for the sea to return," Hasting wrote. Two cruisers and five destroyers came in as close as they could to bombard the port area, and Inchon began to burn. Fighter-bombers from aircraft carriers flew over the city and the roads beyond, ready to attack any attempt by the North Koreans to reinforce their coastal positions. A British war correspondent described how the landing craft milled about, marking time: "There seemed to be no special hurry. We could not go in until the tide was right." Finally, in late afternoon, the first Marines went in on the rising tide, hitting the seawall at 5:31 P.M. and swarming up with ladders and grappling hooks into the city. There were some pockets of stiff resistance, but nothing that could prevent the main body of the assault from landing. Just before dark, as Hastings wrote, "The seal was set upon the American triumph when eight LSTs [Landing Ship Tank vessels] grounded side by side against the seawall on Red Beach. As the tide fell, they remained there, 'dried out,' and from their cavernous holds poured forth a stream of tanks, trucks, jeeps, stores, laying vital flesh upon the bones of the beachhead." The invasion had cost the Marines 20 dead, 1 missing, and 174 wounded, a dramatic victory.

The next day, the Marines pushed inland, capturing a crucial military airfield east of Inchon on September 17. British, South Korean, and other UN forces followed. Seoul, the capital, fell on September 27 as the North Koreans fled in disarray. Intervention by Chinese troops a few months later changed the entire nature of the war, which dragged on for three more years. In the short run, however, the Inchon landing, coordinated with exquisite precision to the movement of the tides, had saved the UN position in Korea and turned the metaphorical tide of the war. And yet, only about half a century earlier, making such precise tidal predictions for a minor port in a foreign country would have been impossible.

The Big Picture

Climate Change and Worldwide Tidal Patterns

Outside a barn-like boat shed overlooking a sheltered saltwater lagoon, a lithe teenager with coffee brown skin axed notches in a long log of breadfruit wood. Another youth used short, controlled swings of his adze to smooth out the rough cuts. They were building a traditional outrigger canoe at Majuro, the most populated atoll in the Marshall Islands of Micronesia.

I had come mainly to watch the annual outrigger sailing races involving canoes from two dozen atolls, each an irregular necklace of coral islands and reefs surrounding a lagoon. I also got the opportunity to sail on one of the spindly but graceful canoes, a seven-metre (twenty-three-foot) long vessel with a hull that tapered to a point at each end and a triangular sail. Two teenagers served as crew, while three fellow passengers and I sat on a platform that extended from the hull to the wooden outrigger. The skipper was a wiry little man named Hanej Helbi. The kids pushed us off as Helbi raised the sail. Instantly we accelerated in the brisk trade wind.

Across the turquoise water our craft skimmed, churning a frothy white wake and going so fast it felt like we were flying. One of the kids stood precariously on the stern, gripping a long steering oar. Occasionally Helbi gestured to us to shift our weight and help keep the boat in proper trim. Like many islanders, he spoke no English. We scooted across the lagoon toward a small island studded with coconut trees, flitting past Taiwanese, Russian, and Panamanian fish processing ships that were anchored among a fleet of sleek tuna seine boats with clipper bows and tiny scouting helicopters.

Along both sides of the forty-kilometre (twenty-five-mile) lagoon, the palm-lined shores faded into the tropical haze. The farthest end was invisible, so low lying that it was over the horizon. And that really struck me. It was my first visit to an atoll, and from what I could see, even the highest land was no more than two or three metres (ten feet or less) above sea level. This is typical of atolls because of how they are formed. As Charles Darwin was the first to explain, each atoll develops from millennia of coral growth around the edges of a volcanic island. At first the coral forms a fringing reef, like the one in Hawaii where I snorkelled over the reef flat. The coral can only grow up to about low-tide sea level. If the volcano gradually subsides, over millennia the corals keep accumulating, remaining at or just below sea level. Eventually, the volcano may disappear entirely, leaving on the surface only a large irregular ring of coral reefs and islands — the atoll. Since the last Ice Age, there was a period of several thousand years when the sea level in the tropical Pacific was about two metres (six and a half feet) higher than it is today. Living coral was able to grow up to that level, and a thin accumulation of soil (from seashells and reef detritus tossed up by waves, bird guano, and organic matter shed by palms and other foliage) added perhaps another metre (three feet) or so. Then sea level dropped, but only slightly, leaving the atoll protruding just above sea level.

So, the low shores of Majuro and the other Marshall Islands are typical of atolls. Some, such as the small island nation of Tuvalu, are even lower lying. In certain places, when storm waves happen to coincide with high tide, they can sweep right across the ring of land. World sea level is expected to rise as much as a metre (three feet) this century due mainly to global warming. Some recent studies indicate that the rise could be even greater. In part, this is due to melting ice sheets in Greenland and

Antarctica, in part to the expansion of ocean water as it becomes slightly warmer. Whatever the cause, large portions of Tuvalu and other very low atolls may simply disappear at high tide.

Tidal Cycle May Contribute to Global Warming

There may be an even more ominous link between the tides and global warming, one with consequences and implications that extend far beyond the fate of sparsely populated atolls in the Pacific. It is possible that a long-term cycle in the tides is influencing the world's climate, leading to greater global warming than anthropogenic (human-caused) factors (such as the increase in atmospheric carbon dioxide and other greenhouse gases) would account for on their own.

This radical idea was first suggested by Charles Keeling, of the Scripps Institution of Oceanography in San Diego, and later elaborated in papers by Keeling and his Scripps colleague Timothy Whorf. (Both have since died.) Keeling was one of the most respected geochemists in the world. He was the person who, in the early 1960s, first documented the steady increase in atmospheric carbon dioxide by measuring it on the peak of the Mauna Loa volcano in Hawaii. (The diagram showing this upward trend over decades came to be known as the Keeling Curve).

In 1995, Keeling and Whorf noticed that periodic changes in surface air temperatures appeared to have a cyclic nature (repeated every six to nine years) that could not be accounted for entirely by cyclical changes in sunspots or other variations in the strength of the Sun's radiation. They suggested that "extreme oceanic tides may produce changes in sea surface temperature at repeat periods, which alternate between approximately one-third and one-half of the lunar nodal cycle of 18.6 years. These alternations … reflect varying slight degrees of misalignment and departures from the closest approach of the Earth with the Moon and Sun at times of extreme tide raising forces." In other words, there were regular cycles of somewhat stronger and somewhat weaker tides. How the changing strength of tides might affect air temperature was that "the dissipation of extreme tides increases vertical mixing of sea water, thereby causing episodic cooling near the sea surface." Cold water from the ocean depths would be brought to the surface to a greater extent in periods when tides were larger than when they were smaller. And they hinted at the next step in their research: "If this mechanism correctly explains near-decadal temperature periodicities, it may also apply to variability in temperature and climate on other time-scales, even millennial and longer."

Over the next few years, Gerald Bond of Columbia University's Lamont-Doherty Observatory (together with several colleagues) looked at evidence of temperature fluctuations recorded in cores from the Greenland ice sheet and from deep sea sediments. He noticed that there was a 1,500- to 1,800-year cycle in which the Earth's atmosphere repeatedly became slightly warmer and then cooler. And this was reflected in human history. In the most recent cycle, for example, Greenland had a relatively mild climate between around 900 and 1200, when the Vikings arrived and initially thrived there. Climatologists call this the Medieval Warm Period, and it was a major blessing for farmers in northern Europe as well, where wine grapes flourished in southern England.

Then conditions deteriorated. Dramatic cooling made it impossible for the Vikings to grow crops to feed their animals. Expanding pack ice made the coastline of Greenland difficult to reach by ship. By around 1350 the more northerly of two Viking colonies in Greenland had perished, and eventually the second one died out as well. Europe was plunged into a cold period called the Little Ice Age, which climaxed around the mid-1500s, when famous paintings by Brueghel show people skating on the canals of Holland, which froze over most winters. In Britain, the Thames froze so solid most winters that merchants set up stalls on the ice and Londoners treated it as a kind of winter carnival. The cold spell extended into the 1800s.

In 2000, Keeling and Whorf picked up on Bond's observation and pointed out that this cycle approximately matched a similar cycle of alternating stronger and weaker tides. They proposed that this tidal cycle of roughly 1,800 years had, not surprisingly, an astronomical cause, namely "almost periodic variations in the strength of the global oceanic tide-raising forces caused by resonances in the periodic motions of the earth and moon." As we have seen with the Saros cycle, there are repeated long-term cycles in the tide-generating forces. The largest tides can be expected to occur when the Moon and Earth are in alignment with the Sun (syzygy) at the same time that each is at its closest distance (in other words, when the Moon is at perigee and the Earth is at perihelion). In addition to these conditions, tides are at their largest when the Moon is at or close to a node, one of the two points in its orbit where it crosses the ecliptic, the plane of the Earth's orbit around the Sun. In the words of Keeling and Whorf, "When an analysis is made to find the times when all four conditions are most closely met, the 1,800-year cycle becomes apparent as a slow progression of solar-lunar declinational differences that coincide with progressive weakening and then strengthening of successive centennial maxima in tidal forces." The strength of the tide-generating forces peaks every 1,800 years, but "progressively less close matching of perigee, node, and perihelion with syzygy occur, on average, at intervals of 360, 180, 90, 18, and 9 years."

The last time the tide-generating forces were at their maximum was around 1425, during the Little Ice Age" in Europe. There have been lesser interim peaks in these forces roughly every 180 years since then, each one a bit weaker than the one before, and after each such peak the strength of the tide-generating forces dropped off quite considerably. So, where does that put us today? The most recent interim peak was in 1974. As Timothy Whorf said in an interview with the Environmental News Network (ENN), "one such episode of cooling during the period 1940–1975, when tidal forcing was stronger, may have temporarily masked the appearance of the greenhouse effect in global temperatures, and contributed to the controversy of whether greenhouse warming was occurring at all."

The next peak of strong tides, a considerably lesser interim one, will not be until 2133. Since 1974, therefore, the tide-generating forces have become weaker and weaker, which means less vigorous vertical mixing of ocean waters. The result, Keeling and Whorf argued, is a trend toward relatively less cooling of the air at the ocean's surface, i.e. global warming. This is on top of any effects that may be caused by human actions, such as increasing carbon dioxide in the atmosphere. And it is a trend that will continue strongly for more than the next few decades. After the peak of 2133, their evidence shows, the weakening trend (encouraging global warming) will become even more pronounced over the following 180 or more years, after which (ignoring shorter cycles of up and down) the curve will level off a bit and begin to rise again. In other words, there will be a trend toward stronger tides and gradual global cooling will set in.

The Keeling-Whorf paper, showing that tides were a "possible cause of rapid climate change," made headlines not only in scientific journals but in the popular media as well. And it was understandable, because the implications are far reaching. If their analysis is right, nothing we do today can prevent increasingly serious global warming over the next few hundred years. A concerted effort to minimize carbon dioxide emissions might ameliorate the impact, but the long-term trend will still be toward ever-warmer temperatures, the melting back of ice sheets, glaciers, and Arctic or Antarctic pack ice, and a drastic rise in sea level. Conditions for certain wildlife, such as polar bears (which depend on long periods of reliable pack ice), will continue to worsen. Permafrost will continue to melt. Island nations such as Tuvalu and the Marshalls, based on low-lying atolls, are simply doomed to face the onslaught of the sea during episodes of high spring tides accompanied by storm surges. So are coastal cities, especially the lowest-lying ones, such as New Orleans. And the impact may arrive much quicker than expected. If he and Whorf are right, Keeling told ENN, and "today's natural warming trend is combined with the greenhouse effect, then we'll soon see the effect of combined warming all over the world."

A heavyweight critique was not long in coming. Their Scripps Institution colleagues Walter Munk and Mathew Dzieciuch (together with Steven Jayne of the University of Colorado) soon published a rejoinder, "Millennial Climate Variability: Is There a Tidal Connection?" If Keeling was, until his death, one of the most respected geophysicists in the world, Munk has long been among the world's leading oceanographers and tidal specialists. Munk and his co-authors argued that the astronomical influences on tidal mixing (peaks and troughs in the strength of the tide-generating forces) were both short-lived (when they occurred) and most likely of insufficient amplitude to account for the observed millennial variability in the climate. They did not deny that the variability exists, but sought the primary explanation elsewhere. One possibility, and the one they thought most probable, was instabilities (which were admittedly not yet identified) in the dynamic interactions between the ocean and the atmosphere. Another was a variation (likewise not yet discovered) in the amount of radiation reaching Earth from the Sun over a one-thousand to two-thousand-year cycle. Either alternative required further research. They conceded, however, that a harmonic cycle in the orbits of Sun, Earth, and Moon "cannot be ruled out by any evidence known to us," and that if any such cycle *is* a factor in climate change, the cycle identified by Keeling and Whorf "is the most likely candidate."

In short, the jury is still out, which is awkward considering the weighty (and potentially controversial) practical and political implications. Because if cumulative human activity is not the sole (or perhaps even the most important) contributing factor to global warming in our time, then cutting back on greenhouse gas emissions cannot entirely solve the problem, no matter what resources or changes in our technological practices and economic activities are devoted to the effort.

Keeling and Whorf did not emphasize this sticky issue in their public statements, and Keeling himself was really the father of climate change based on greenhouse gases. But if a natural cycle inherent in the astronomical geometry of the Sun-Earth-Moon system is a major cause of global warming, then significant climate change over the next few hundred years is inevitable. In that case, the most sensible strategy for the human race may involve focusing our efforts and resources on adapting to the unavoidable consequences. This could include building dikes or raising existing ones, relocating coastal population centres, helping endangered wildlife to find new habitat, encouraging farmers to diversify or grow different crops, etc. And it might make less sense to pursue expensive crash programs aimed at reducing greenhouse gas emissions, which may be doomed to failure. Or perhaps some balance of effort combining the two approaches may be required. But weighing the alternatives would surely be worth considering.

The implications are dire, and they run counter to the reports and agenda of the UN International Commission on Climate Change with its primary focus on reducing greenhouse gas emissions. Possibly for this reason, after an initial flurry of attention and media coverage, the environmental movement has almost entirely ignored the key Keeling-Whorf papers and several follow-up papers and presentations. Most recent books on climate change do not mention the tides as a possible factor at all. Perhaps leading environmentalists simply do not want to hear that the tides could be causing global warming. They may see this as lending support to the climate change skeptics, who point to natural cycles and play down or deny the anthropogenic causes of any observed rise in temperatures. Furthermore, environmentalists tend to be activists committed to analyzing a problem, finding the cause, and setting out to solve the problem by removing the cause or reducing its effects. But if Keeling and Whorf are right, there are strict limits on what the human race can do about global warming.

Meanwhile, we can continue to study the operation of the tides, although that pursuit, too, is now nearing its practical end.

Rollin Harris Maps World Tidal Systems

Our ability to predict the tides at specific ports made considerable progress in the latter part of the nineteenth century, culminating in William Thomson's harmonic analysis and his invention of the first rudimentary tide prediction machine. By the 1880s and 1890s, using a more advanced tide prediction machine developed by American mathematician William Ferrel, it was possible to begin generating reasonably accurate tide tables for most major ports, based on observations lasting at most only a year or two. But the "big picture" of how the tides worked across entire ocean basins was still quite murky. And there was little agreement among scientists on some of the key principles involved. The progressive wave theory had been challenged by other approaches based on the concept of resonance, but many theorists, including such prestigious scientists as George Darwin, clung to it. And no one could adequately explain the overall pattern of tides around the entire world. Some basic insights were apparently missing.

One region where late–nineteenth-century tidal science was a particularly woeful failure was around the shores of Australia. The progressive wave theory envisaged a tidal wave travelling from east to west in the great Southern Ocean (the only place where a wave could move unobstructed around the world) and propagating northward successively into the Indian, Atlantic, and Pacific Oceans. As we have seen, the size of tides and their timing in the Atlantic could not be explained by a simple progressive wave propagating from south to north. Nevertheless, the progressive wave approach was not readily abandoned. As late as 1926, a map issued by the U.S. Navy's Hydrographic Office included a series of co-tidal lines supposedly representing all points at which high water arrived at the same hour. For Australia, it showed the tidal wave approaching the east coast of Australia from the Pacific, moving from east to west along the southern shore of that continent, and then flowing northwestward across the Indian Ocean. But this was not based on empirical evidence.

As Sir Robert Chapman, a professor of mathematics, mechanics, and engineering at the University of Adelaide, commented in 1938, "It is obvious that in the making of such a map of co-tidal lines the imagination of the author has been brought into play quite considerably, because we have no observations of the rise and fall of the water at points far

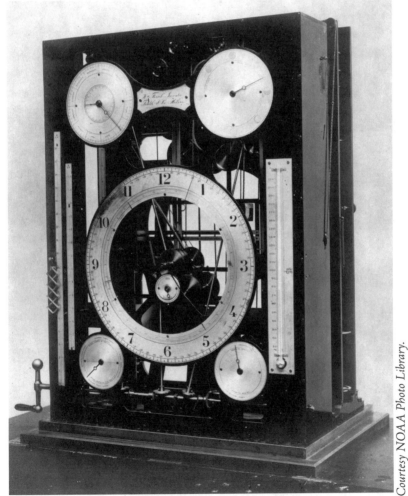

William Ferrel's tide prediction machine, front view.

Courtesy NOAA Photo Library.

William Ferrel's tide prediction machine, back view.

out from land, and our actual observations, upon which the map of co-tidal lines is based, are confined to places on the shores of the continents and to islands." Chapman added that, other than Tasmania, "there are no islands in the ocean to the south of Australia, and it follows that the shape of the co-tidal lines in that region in this map must be determined by the progressive wave theory which the author has in his mind rather than by actual observation."

When he looked at places where accurate tidal measurements *had* been made, Chapman found things that made no sense in light of that theory. "If ... the tides along the south coast of Australia ... are due to a tidal wave moving from east to west, how is it that from Cape Howe [near the southeastern tip of Australia] to the Head of the Great Australian Bight, more than halfway along, we have a mean spring range of tide running from 5 to 6 feet [1.5 to 1.8 metres], whereas from there to Cape Leeuwin [near the southwestern tip] the range is only about 2.5 feet [0.76 metres]. It cannot be explained either by a variation in the depth of the ocean or by a change in its width." The situation was even more bizarre along the west coast of Australia, where going "from south to north, tidal range at [spring tides] from Cape Leeuwin up as far as Dirk Hartog Island is less than 3 feet [under one metre], but from there it increases rapidly until at Port Hedland it is 19 feet 3 inches [5.9 metres]. The progressive wave theory alone does not give us any reasonable explanation of facts like these."

The challenge to offer a better account of world tidal patterns was taken up in the first years of the twentieth century by an American mathematician educated at Cornell University, Dr. Rollin Harris, whose career was devoted to computing tides for the U.S. Coast and Geodetic Survey. Harris hearkened back to the mid-nineteenth century ideas of Fitzroy and Airy — that ocean basins might oscillate, or rock back and forth in the form of standing waves — and to Whewell, who had envisaged rotational systems of waves moving around a point of no tide. Instead of a progressive wave moving around the Southern Ocean, Harris saw each ocean as made up of one or more largely separate great basins of water that are repeatedly disturbed by the periodic tide-generating forces. Because of the different depths and dimensions of the ocean basins, each will have a somewhat different natural period of oscillation and will oscillate more or less to keep time with a different combination of tide-generating forces, according to their periods. Some ocean basins may be more inclined to oscillate in time to lunar forces, for example, others to solar forces. The amplitude of the oscillations may differ greatly between ocean basins, and may also differ from what we would expect simply by looking at the simple astronomy of Newton's tidal theory.

Returning to the Australian example, Chapman pointed out that "the tide-producing forces due to the moon are about 2.3 times as great as those due to the sun, but we do not find that the lunar semi-diurnal tide is everywhere 2.3 times as great as the solar semi-diurnal tide. There are places around the coast of Australia where the solar tide is just as big as the lunar tide, and other places where the lunar tide is five or six or even, as on the New Zealand coast, ten times as big as the solar." He explained, "The most reasonable explanation of such effects ... is that they are due to the selective resonance of some adjoining body of water. If, for example, the solar semi-diurnal tide is much greater than we should expect, in comparison with the semi-diurnal tide due to the moon, the probable reason is that there is an adjacent basin of water that has a natural period of oscillation of just about twelve solar hours, which harmonizes with the period of the sun's tide-producing forces."

Rollin Harris looked at the world's oceans and seas and, based on their dimensions and what was (very imperfectly) known in the early twentieth century about their recorded depths, he divided them into areas that, he calculated, should oscillate in rhythm with one or another of the major tide-generating forces. He also adopted Whewell's concept of rotational systems circling around points of no-tide, and he coined the term *amphidromic system* (from the Greek words *amphi*, or "around," and *dromos*, or "running"), with the amphidromic point being at the centre. He drew a series of theoretical maps representing how he thought the major ocean basins should respond to the dominant tidal harmonic, the lunar semi-diurnal or M 2 tide.

When Harris published his most advanced co-tidal map for the south coast of Australia, it showed the tide-wave approaching the coast from the south, not propagating from east to west (as the orthodox map continued to show as late as 1926), and it fit the observed data much better. According to tide-gauge measurements at Port Fairy and Streaky Bay, two ports on Australia's southern coast separated by about eight degrees of longitude — some 750 kilometres (460 miles) in an east-west direction — high tide arrived at both within eleven minutes of each other. No progressive wave could move from east to west that fast.

The west coast of Australia, he suspected, was subject to a tidal regime that prevailed in a huge swath of the Indian ocean between Australia and Madagascar, not to tides coming westward and around from the Southern Ocean. Harris also surmised that there was a major amphidromic system roughly in the middle of the North Atlantic, one in the northern Indian Ocean, and several in the Pacific. (In fact, there are now known to be several other major amphidromic systems as well. See Figures 12.1 and 12.2 for recent maps based on satellite observations.) Most of the amphidromic systems located north of the equator rotated in a counter-clockwise direction, while most in the southern hemisphere rotated clockwise. Harris's co-tidal map of the North Atlantic correctly showed an amphidromic system centred in mid-ocean, with the co-tidal lines sweeping around counter-clockwise. High tide in Europe would occur first at Gibraltar, then northern France about an hour later, followed by Ireland an hour after that and Iceland about two hours later. Similarly for the North Pacific, the amphidromic system west of the North American coast would bring high tides moving northward at approximately hourly increments starting with southern Mexico, then northern Mexico, southern California, northern California, British Columbia, and finally Alaska.

Modern studies have shown that in some cases, where adjacent rotating amphidromic systems meet they can interact with each other, much like a set of meshing gears that rotate in opposite directions. The dominant waves from one system can cause the waves of the other system to rotate in a direction that is counter to what would be expected if the second system were rotating solely in accordance with the tide-generating forces, the shape of its own ocean basin, and the coriolis effect. For this reason, amphidromic systems centred on Madagascar and in the mid-Pacific just south of the equator rotate in the "wrong" direction.

Harris also noted that marginal basins, such as the North Sea, are too small (and in many cases too shallow) to have tides of their own produced directly in resonance with the tide-generating forces. Instead, progressive waves from large ocean basins (from the Atlantic, in the case of the North Sea) propagate into them and form smaller amphidromic systems, as Whewell had discovered for the North Sea. Although Harris's overview of worldwide tides has been modified and improved upon during the past century, his big picture of ocean basins dominated by discrete amphidromic systems comes close to representing how contemporary tidal science views the operation of the tides.

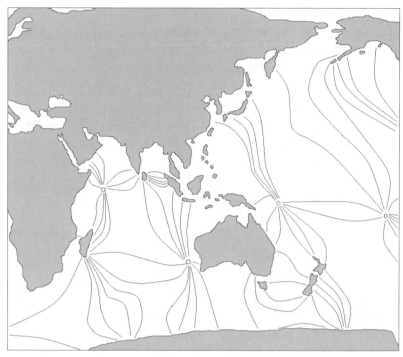

Figure 12.1: Amphidromic systems, eastern hemisphere.

The Refinement of Modern Tide Prediction

During the first half of the twentieth century, more advanced tide-prediction machines based on up to sixty-two harmonic components led to predictions that were generally accurate enough for nearly all the practical purposes demanded by mariners, port authorities, and the like. Since then, computers have made the process quicker and easier, but only slightly more precise. The most significant advances in tidal science since the mid-twentieth century have come with the introduction of highly accurate radar altimetry from satellites, which makes it possible to measure the height of the sea and variations in that height (sea surface topography) from space with an accuracy of several centimetres. This has filled all the former gaps in sea level measurement.

At one time, the height of the tide (and therefore the movement of tide-waves) could only be gauged along continental shorelines or at scattered mid-ocean islands. This made it impossible to be certain of tidal

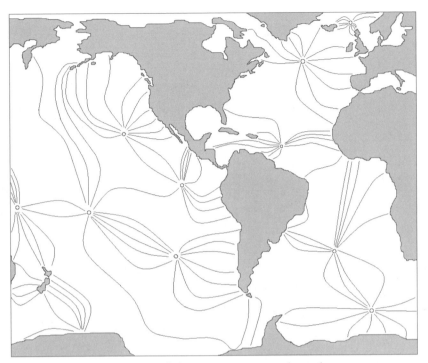

Figure 12.2: Amphidromic systems, western hemisphere.

patterns across much of the world's oceans, because there were huge areas (such as the Arctic Ocean, the southern Indian Ocean, and most of the seas around Antarctica) where there were either few islands or none at all. Today, tide-waves can be mapped accurately in all oceans, and all the major harmonic components contributing to them have been isolated and analyzed. As British tidal scientist David Cartwright comments, "Space technology has thus come near to bringing about the culmination of an era of research into mapping the tides which started in 1833 with the initial speculations of John Lubbock and William Whewell." Cartwright adds, "Most of the original mainstream problems of the tides are now part of the history of science.... The accuracy of tide-tables for most ports is limited by the less predictable effects of weather [rather than by] astronomical effects." The big picture of tidal generation and movements is, therefore, essentially a completed tableau.

Turning the Tide

Tapping the Energy of the Oceans

The oddly shaped but vaguely futuristic building, with lots of glass and huge external ductwork, was perched on a long causeway across the mouth of the Annapolis River in Nova Scotia, just upstream from where it flows into the Bay of Fundy. Seeing the Annapolis Royal tidal power plant in action, generating up to twenty megawatts of electricity from a clean, renewable energy source, was something I had eagerly looked forward to during my tour of the bay. I timed my visit to the facility to coincide with a period of ebbing tide, when water impounded behind the causeway in the river's estuary would be rushing out through the single giant turbine and cranking out power.

Alas, it was not to be. When I met chief technician Tom Foley, who has personally operated the plant since it first opened in 1984, he broke the news to me. The facility was down for repairs, and Foley was having a bad hair day trying to isolate the difficulty. Still, after showing me the control room with its banks of dials and switches, he took the time to escort me by elevator down into the bowels of the plant, where a steel

Photo by Tom Koppel.

The Annapolis Tidal Generating Station is North America's only tidal power plant. Located in Annapolis Royal, Nova Scotia, the plant produces 20 megawatts of electricity for Nova Scotia Power Inc.'s electric grid.

catwalk afforded a view of the silent, non-functioning turbine. It was several storeys high and equally wide and was situated well below the level of water in the estuary. Foley pointed to the various parts of the turbine and explained their functions. "There's a problem with the stator," he said, indicating the huge, ring-shaped set of wire coils that, together with the rotor and its magnetic poles, generates the electricity when fast-flowing water makes the rotor spin. "We don't know what's the matter with it. That's the problem."

He assured me that days when the plant was out of commission were very rare and that tidal power is actually more reliable in that sense than any of the alternatives, such as thermal or nuclear power plants. "Our availability rate has been in the vicinity of 95 to 98 percent since we started up," he said with obvious pride. But he was also frank about the disadvantages of depending on energy from the tides. "The biggest problem with tidal power is that it's not there 24/7." Because the Annapolis Royal plant generates only on the two daily ebb tides, it produces power for a total of just eleven hours each day. So its contribution to the energy supply of the Nova Scotia Power Inc. grid is predictable, because the tides are predictable, but only intermittent.

Harnessing the energy of the tides for human needs is not a new idea. In medieval Britain, France, and Spain, several hundred tidal mills were in use, mainly to grind grain into flour. One British mill operated commercially for eight hundred years, until 1957, and is now a museum. A high dam-like barrier, or barrage, was built across the narrow entrance to a tidal inlet. The flooding tide was allowed in through a sluice gate to raise the level of water behind the barrage. Then, as the tide ebbed, it flowed through a water wheel, and the rotational motion was used in the same way that the wind mills of that era exploited wind power. The Annapolis Royal power plant is based on the same principle, except that it generates electricity with a turbine, much the way a standard hydro-electric dam does.

Considering the vast amount of energy embodied in the rise and fall of the tides, exploiting tidal power seems like an obvious solution to many of the world's energy needs. And yet Annapolis Royal is the only tidal power plant in North America. It produces enough electricity for no more than about three thousand homes, yet that makes it the second largest tidal power station in the entire world. There is a much smaller one in northern Russia, and one in France whose output is more than ten times as large. That is the La Rance power station, which was completed in 1967 on the coast of Brittany, not far from the abbey of Mont Saint-Michel. A barrage seals off the estuary of the Rance River. Twenty-four large turbines set into the base of the barrage operate during both the flood and ebb cycles, so they produce power for a much larger portion of the day than the ebb-only system at Annapolis Royal. At peak output La Rance generates a total of 240 megawatts, and the installation certainly looks impressive, with its 330-metre (1,083-foot) long barrage and water gushing out through the turbines, much like a sizeable hydroelectric dam. (There is a lock at one end of the barrage for small vessels to pass through between the sea and the river.) The power production, however, is still only intermittent and is considerably less than most commercial nuclear, oil, gas, or coal-fired power plants, and much less than large hydroelectric dams as well.

Consider the Hoover Dam on the Colorado River, for example, which was completed in 1936. It is 379 metres (1,244 feet) long, or only

France's La Rance tidal power station.

slightly longer than the La Rance barrage, but it generates a peak output of 2,074 megawatts, or nearly nine times as much as La Rance. The reason it can generate so much more power is that it is located in a steep, mountainous region and is 221 metres (726 feet) high. Water from the huge lake created behind the dam falls a long distance, with great speed and kinetic energy, before racing through the generating turbines. By comparison, the La Rance barrage is 13 metres (43 feet) high. The tidal range in the Rance estuary averages 8 metres (26 feet) and occasionally peaks at 13.5 metres (44 feet). Each unit of water flowing through the Rance turbines, therefore, has much less energy to harness than in the case of a high hydroelectric dam.

Nevertheless, because such a large volume of water flows in and out of certain bays and estuaries, there is enormous potential for barrages to generate tidal power at many locations in the world where the tidal range is large: on the northern coast of France and at several places on the coast of Britain; on Canada's Bay of Fundy and on the southern and western coasts of Korea. Even where the tidal range is not quite as large, swift tidal currents (at the mouths of inlets or in narrow channels between islands) can also, in principle, be harnessed for energy production. So why is the world's largest tidal power plant so modest in size? And why has no new major tidal plant come on stream in recent years?

The reasons relate mainly to comparative costs and environmental impact.

Take costs, for example. The Annapolis Royal plant was built as a pilot project to study the potential for much larger tidal power projects that had been proposed in the 1960s and 1970s for the Bay of Fundy. The largest of these would have involved building barrages across Cobequid Bay on the Minas Basin in Nova Scotia and across Shepody Bay on the New Brunswick side of the Bay. The larger of these, on the Minas Basin, where the maximum tidal range is around 16.4 metres (54 feet), would have produced a peak capacity of 3,200 megawatts. This is more than at Hoover dam and over thirteen times as much as at La Rance. But the construction cost of the barrage would have been extremely high, and the power output would have greatly exceeded the energy needs of that largely rural region. Moreover, even if power generation took place on both the flood and ebb, there would still be times when backup power (presumably from large thermal plants) was required. Yet, much of the time those plants would be idle. On purely technical grounds, the project might still have been justifiable. Some of the power could have been fed into the continental energy grid to contribute to the electricity supply in Boston and elsewhere in New England, thereby mitigating the downside of tidal energy's intermittent nature. Nevertheless, on a cost basis alone, the prospects were daunting.

"I would have loved to see those developments, which were much larger than this one, go ahead," said Tom Foley. "But back in 1984, in order for [Nova Scotia Power] to build any of the other sites, you had to have the price of oil at over [US]\$60 a barrel. Now, what was the price of oil in 1984? Do you remember? At one point it was \$48 a barrel. That was the high, in those dollars. Today, I'll bet you, it would have to be over \$100 a barrel to make tidal feasible."

Rising oil and other energy prices over the last decade or so, combined with the belief that world fossil fuel availability (other than coal) has peaked and begun to decline, have led to renewed interest in tidal energy. Concern about global climate change has also focused interest on *any* energy technology that does not produce greenhouse gases. Tidal power, therefore, may play a larger role in the future mix of renewable energy sources.

But there is another key consideration that will affect what kind of power projects are developed: environmental impact.

Building a long barrage across the mouth of a bay or estuary inevitably obstructs the natural flow of water. One result is the tendency for silt to collect behind the barrage in the reservoir created by it. Another is the impact on migrating fish and other marine species. The silt can be dredged out, but that is expensive and an intrusion on estuarine marine life. And even though the tide still flushes in and out through the turbines of the barrage (and at times through additional sluice gates that can be built into the system), the natural conditions are significantly altered. At La Rance, for example, sand eels and plaice have disappeared. Fortunately, the Rance is not a really major river.

La Rance was originally conceived, however, as just a pilot project for a truly colossal tidal power plant built into a thirty-five-kilometre (twenty-two-mile) long barrage that would have enclosed the entire Baie de Mont Saint-Michel and provided much of France's power needs. The French eventually opted for nuclear power instead, becoming the world's most nuclear-dependent country. But the environmental consequences of the Rance facility hint at the impact the giant project initially envisaged would have had on the marine life and resources of the northern French coast. Not surprisingly, although they are attracted by the prospect of clean, renewable energy, environmental groups are opposed to most tidal power projects that rely on barrages. This has been apparent in recent years as tidal projects have come back onto the energy agenda.

At Sihwa Lake on the western coast of South Korea, a 260-megawatt tidal power plant is being built at a cost of around US$250 million. When completed in 2009, its peak output will make it comparable to La Rance, but in other ways it is quite different. Sihwa Lake is one of several artificial lakes along that coast that were created as part of projects to gain new industrial and agricultural land from tidal wetlands by building long barrages across the wide mouths of rivers. Since the thirteen-kilometre (eight-mile) long barrage at Sihwa Lake was completed in the early 1990s, pollution from some of the industrial facilities has degraded the water quality in the lake, leading to remediation in the form of using sea water to flush the lake on a regular basis. The average tidal range there is 5.6 metres (18.4 feet), which means that at high tide water in the open sea is higher than water in the lake. The power plant is being built into part of the already existing barrage, which drastically reduces the cost compared to having to build an entirely new barrage. However, it will mainly generate power on the ebbing tide following the peak of high tide, as water flows from the sea through the turbines and into the lake. (Once the level inside the lake is high enough, and the sea level is low enough at the end of the ebb cycle, some of that water will then be allowed to flow out to sea through sluice gates in the barrage.)

Environmentalists have never liked anything about the impact of creating Sihwa Lake. The English-language *Korean Times* calls Sihwa Lake a "failed reclamation project." The Korean Wetlands Alliance says "the once large-scale tidal flats ... have been destroyed [leading to] the destruction of local communities and fisheries, and habitat conditions for migratory birds." Regularly flushing the lake with sea water has been the only way to mitigate the damage done by building the original barrage, and it apparently has worked to a degree. What once was termed a "dead sea" is now a place where many wildfowl overwinter. Building a power plant to exploit the flushing process may make sense, but environmentalists would prefer eliminating the sources of industrial pollution, rather than flushing it out to sea. Yet the barrage is there and huge investments have been made on the land created by it. The power project is going ahead, and similar ones are being considered for other sites along the same stretch of coast.

The granddaddy of all tidal barrage power installations — if it is ever built — would be on the estuary of the Severn River between Wales and Somerset in western Britain. First envisaged in 1923, it would take advantage of the area's peak tidal range of over fifteen metres (forty-nine feet), which is second only to the Bay of Fundy and Ungava Bay. A proposal developed in 1989 (and still being considered) foresees a barrage spanning almost sixteen kilometres (ten miles) with 216 turbines generating 8,640 megawatts, or thirty-six times the output of La Rance, and providing some 6 or 7 percent of Britain's total power needs. It would take fourteen years to build at an estimated cost of around US$28 billion in today's dollars. One side benefit is that a road built across the barrage would provide a shortcut between southwestern England and South Wales.

The proposal lay dormant for well over a decade. Britain had North Sea oil and gas, plus many nuclear plants, and awareness of global climate change grew only gradually. Then, in early 2006, the British government was working on a review of the U.K.'s energy policy and long-term needs. There were expectations that this might result in recommending a new generation of nuclear power stations. The Secretary of State for Wales, Peter Hain, one of the strongest critics of nuclear power in the cabinet of Prime Minister Tony Blair, came out in favour of revisiting the Severn proposal as an alternative to nuclear. Rhodri Morgan, the Labour Party's leader in the Welsh Assembly, agreed, and the British media reported that Prime Minister Blair also supported taking a serious new look at the idea.

The reaction from environmental groups was swift and negative. Gordon James, a spokesperson for the Welsh branch of Friends of the Earth, called it a "kneejerk reaction" to the prospects of nuclear power and said, "We don't need nuclear power and we don't need the barrage. The existing proposals for wind power on land and offshore together with wave power could generate up to a quarter of Wales's energy needs." Morgan Parry, head of the Welsh branch of the World Wildlife Fund, said "constructing a ten-mile concrete energy dinosaur will cause irreversible damage to Wales and England's most important estuaries." The left-leaning *Guardian* daily reported that Tim Stowe, Welsh director of the Royal Society for the Protection of Birds, foresaw harm to "habitat for 65,000 waders, ducks and other birds as well as many species of fish and invertebrates," along with impact on several rivers that feed into the Severn. It would also ruin the Severn tidal bore, "the extraordinary wave which periodically rushes upriver and attracts surfers from across the world." The Royal Society, Stowe added, supports renewable energy, but "risking irreplaceable wildlife sites for the sake of energy generation is not a sustainable option."

It remains to be seen how this issue will be resolved, but one immediate result was enhanced support in the media and from some environmental groups for a more modest and somewhat different type of tidal energy concept on Britain's coasts.

Artificial Tidal Lagoons

The main problem with barrages is that they seal off the entrances to estuaries and inhibit the natural movement of water and marine species into and out of estuaries and tidal wetlands. But what if it were possible to harness the rise and fall of the tides without obstructing these valuable and vulnerable ecosystems? One solution lies in the concept of building artificial, free-standing tidal lagoons. These would be huge, roughly circular enclosures with outer walls made of boulders, rubble, and other relatively inexpensive materials, a bit like the construction of harbour breakwaters. They would be situated offshore just beyond the low tide line on broad, shallow, gently sloping tidal flats in places with large tidal ranges. Sluice gates and power-generating turbines would be incorporated into a portion of their outer walls. Starting with an empty lagoon, when the tide in the outside sea rose and approached the high water mark, the gates would open and water would begin to flow into the lagoon through the turbines, generating electricity. This would eventually fill the lagoon. Then, as the tide in the open sea fell, water would flow out of the lagoon through the turbines throughout the latter part of the ebb cycle and during the period of slack tide, which would also generate electricity. This generation cycle is not very different from that of a barrage system like La Rance.

The beauty of tidal lagoons, however, is that they can be situated in places where they do not obstruct the flow of water in and out of sensitive and wildlife-rich estuaries. They would be built, instead, just off the coast on shallow areas of less environmental value and vulnerability. Among the projects under serious consideration is one to build a US$75 million pilot project that would impound roughly five square kilometres (two square miles) of sea water in Swansea Bay, south Wales, on the Bristol Channel west of the Severn. The average tidal range there is about nine metres (thirty feet). This is much more modest than the kind of mega-scheme envisaged in the Severn Barrage project. Proposed by a British-based company called Tidal Electric, which holds a patent on the concept, this first lagoon would extract a peak capacity of about thirty megawatts, or 50 percent more than the Annapolis Royal plant in Nova Scotia. Unlike the Canadian facility, it would generate almost equally well on both the flood and ebb tides. And there are several other sites, such as on the Severn estuary itself, with its average tidal range of around twelve metres (thirty-nine feet), where much larger tidal lagoons could be built (without blocking much of the estuary) and where the electrical output would be far greater. Proponents also argue that the walls of the offshore lagoons would be ideal places to site wind turbines, and the same transmission lines and corridors could be used.

Some environmental groups in Britain have come out in support of tidal lagoons as a more reasonable, and greatly preferable, alternative to barrage-based tidal power. Friends of the Earth Cymru (in Wales) issued a position paper in 2004 titled "A Severn Barrage or Tidal Lagoons." The document relied almost entirely on technical data and projections provided by the Tidal Electric company, which has an interest in making a strong case for lagoons, so its findings should be viewed with a measure of skepticism. It argued that for the Severn Estuary, "several large lagoons could be built over time to impound as much as 115 square miles [298 square kilometres] of the Estuary.... Even then the lagoons would cover 70 square miles [181 square kilometres] less than the 185 square miles [479 square kilometres] the Barrage would impound. Yet this area of lagoons would capture about 26–41% more of the Estuary's tidal energy than the Barrage." It continued, "This is because the lagoons can generate electricity on both the ebb and flood tides while the Barrage is limited to generating mostly on the ebb tide to reduce silting.... Overall, lagoons would generate just over twice as much electricity per square mile impounded than the Barrage."

As for cost, the paper estimated that electricity from lagoons in the Severn Estuary would be "highly cost-competitive rivalling the forecast price of offshore windfarms and the cost of gas generated electricity. Indeed, gas prices are now rising due to increasing global demand and the UK is likely to be a net importer by 2006.... Tidal Electric says that lagoon schemes would produce power at about 2 pence per Unit (kW hour) [US$0.04] much cheaper than the barrage."

A key concern of Friends of the Earth is, understandably, the environmental effects. "Generally speaking," the report went on, "minimizing the impoundment of the particularly rich inter-tidal areas and tributary rivers, is better. In this respect the lagoons appear to be far preferable to the Barrage. The nearest coast-facing lagoon walls would be located around the low-water mark, typically up to one mile from the coast, and would avoid the inter tidal areas." On balance, therefore, "on the basis of this preliminary analysis and comparison tidal lagoons could provide a major source of safe, clean, regionally generated renewable electricity. Lagoons also appear to offer numerous significant economic and environmental advantages over a Severn Barrage." The organization called on the Welsh Assembly to support further detailed studies of tidal lagoon schemes for the Severn.

Two years later, when renewed political interest was shown in the Severn Barrage scheme in early 2006, the World Wildlife Fund of Wales also urged consideration of alternative tidal power projects on the Severn, "such as stand-alone tidal generators, tidal fences," which would harness tidal currents, "and further research into tidal lagoons." Because they are more modest in size and environmentally more acceptable than the huge barrage concept, it appears likely that at least some tidal lagoons will be built on the south coast of Wales within the next decade or so.

There are other places, as well, where they may be suitable, including on the north coast of Wales, in the Bay of Fundy, and on the coasts of Bangladesh and the state of Gujarat, India. In 2004 the governor of Liaoning Province in China signed a preliminary agreement with Tidal Electric for the technology to build a three-hundred-megawatt tidal lagoon power station near the mouth of the Yalu River. The next step would be engineering feasibility studies. However, not all shorelines where the tidal range is large have the kind of wide, gently sloping offshore profile that lends itself to this type of construction. So far, no tidal lagoon scheme has yet moved beyond the talk and study stage to actual project funding and construction.

Exploiting Swift Currents

An entirely different way of harnessing tidal energy (although as yet untried on any significant scale) is to generate electricity from swift-flowing tidal currents, especially in narrow straits and channels, between islands and at the mouths of inlets and river estuaries. The most common concept is for turbines with large blades to be placed underwater, where the tidal current will turn the turbine on both the flood and ebb cycles. If it is a navigable channel, they must be fixed or anchored deep enough so they do not obstruct shipping. Most existing turbine designs work well if the current speed is four knots or higher. In modified form, this type of energy production is already employed in fresh water on some fast-flowing rivers, although in such situations (called run-of-the-river power) the turbines operate only in one direction. As with other tidal power systems, there would be periods when no power is produced, such as when the current in one direction weakens and goes slack before turning and eventually flowing with strength in the opposite direction.

Among the advantages of tidal current power (as opposed to barrages and even tidal lagoons) is that, because no water is impounded, there should be considerably less impact on the environment. The migration of fish is not obstructed, there is no siltation of estuaries or rise in water temperature (as may occur with an impoundment), and there is only a very small footprint" where the installation is attached to the sea floor. The system can be expected to work well on both flooding and ebbing tides, since in most tidal channels the currents generated are of equal, or very nearly equal, speed in both directions. Finally, there are many places in the world that have swift tidal currents, but not the geography that lends itself to either barrages or tidal lagoons.

The coast of British Columbia is a prime example. The only places where tidal barrages might work well are at the mouths of large rivers (the Skeena, for example) that are rich in migratory fish species, especially salmon. It would be environmentally and politically unacceptable to obstruct these rivers. There are few, if any, places with broad tidal flats and sufficient tidal range to lend themselves to free-standing tidal lagoons. However, coastal B.C. has dozens of channels with extremely fast-flowing tidal currents, many of them peaking at eight knots or more. Some are among the swiftest such currents in the world. Tidal turbines, or other technologies that harness tidal currents, could be placed in many of these channels at depths or locations where they would not obstruct navigation. In fact, much of the Inside Passage, from the Strait of Georgia to southeastern Alaska, where the cruise ships wend their way through deep, narrow channels between steep mountain slopes, has great potential for tidal current power.

Many other regions in the world offer good prospects as well. Around Britain, these include parts of the English Channel, the Irish Sea, and especially the Pentland Firth north of Scotland, which has about 40 percent of Britain's potential tidal current resource. In Scandinavia, there are the Skagerrak and Kattegat. In the Bay of Fundy, there is the channel where the tide surges into Minas Basin past Cape Split. There are also several good locations in northwestern and western Australia, on the coast of Brittany in France, in the Strait of Magellan, and in straits between islands in the Philippines.

How much power could be generated from these tidal currents? To give a sense of the scale, a 2001 study by Britain's Parliament places the U.K.'s total electricity consumption at 330 terawatt hours (each equal to one trillion kilowatt hours) per year and says that exploitation of the ten most promising U.K. sites for tidal energy of all kinds could generate a total of 36 terawatt hours per year. A 2003 Belgian academic study estimates that if all the best 106 tidal current locations in Europe were put into service, they could provide 48 terawatt hours per year, about half of that from British waters. By way of comparison, if the three largest proposed barrage projects on the Bay of Fundy were built, they would generate about 16 terawatt hours per year. And a British Columbia Hydro company study in 2002 estimated that the realistic potential for tidal current energy on the B.C. coast is about 20 terawatt hours.

Despite the considerable promise of tidal current power, only one minute pilot project (in the Norwegian Arctic) exists so far. It generates enough power for just thirty homes. But projects employing different kinds of tidal current technology are now on the drawing board in several countries. Given the environmental downside of barrages and the limited number of places where tidal lagoons are suitable, tidal current power seems likely to be the "wave" of the tidal energy future.

A number of turbine designs and turbine placement configurations have emerged in recent years. Most look a bit like wind turbines. They can swivel to face into the current, whether it is ebbing or flooding, and have vertical blades ten to twenty metres (thirty-three to sixty-six feet) in diameter that rotate on a horizontal shaft, which in turn drives a generation unit behind the blades. The base of each unit can be rigidly fixed (in steel and concrete) to the sea bottom, or it can be anchored to the bottom with a set of strong cables and hang or float suspended at an adjustable height between the sea bottom and the surface. The choice would depend mainly on the depth of the water, nature of the seabed in any particular location, and navigational use of the channel.

Most designs incorporate some way to detach and bring the turbine to the surface for maintenance. This obviates the risk and expense of using divers for such tasks. The turbines would be staggered and spaced to minimize their interference with the flow of water around neighbouring turbines and to maximize their total output. (Windfarms also have to be designed with the turbines properly spaced and staggered.)

Although proponents of tidal current energy extol their minimal environmental impact, some environmentalists and biologists worry that the rotating, windmill-like blades could hurt fish and marine mammals, especially large ones such as whales. One Canadian company, Vancouver-based Blue Energy, has developed a different kind of turbine that, the firm claims, solves this problem. Its Davis turbine employs a large-diameter wheel-like steel framework that holds narrow vertical hydrofoil blades on its outer rim. The wheel rotates in the horizontal plane (around a vertical axis), and the hydrofoils have a shape akin to airplane wings, which creates "lift" and allows the rim of the rotating structure to turn faster than the speed of the current. Yet the speed is still slow enough that it would do no harm to ordinary migrating fish. Larger marine mammals, such as whales and dolphins, can be kept out of the rotating structure by protective large-mesh fencing.

A huge system was designed by Blue Energy in 2000–2001 to span the four-kilometre (two-and-a-half-mile) wide San Bernardino Strait between two islands in the Philippines, where tidal currents run at up to eight knots. At an estimated cost of US$2.8 billion, it would have used 274 Davis turbines, each generating from 7 to 14 megawatts, for a total peak output of 2,200 megawatts. This was intended to be just phase one of a larger proposed project that would have spanned three adjacent straits as well for a total peak output of 25,000 megawatts. The project foundered due to political instability and a corruption scandal in the Philippine government.

Alternatives to Turbines

Not all systems for harnessing the tides involve rotating blades or wheels. One type of British device, called a Stingray, is designed to work in currents as slow as three knots. It employs a fifteen-metre (forty-nine-foot) long wing-like hydroplane mounted horizontally at the end of an arm hinged to an upright post that is firmly mounted on the sea floor. The tidal current acts on the hydroplane to make it oscillate up and down in a sort of nodding motion. (The angle of the hydroplane is set differently on the upstroke and downstroke to catch the current.) This activates hydraulic cylinders, which pump high-pressure hydraulic fluid, and that, in turn, drives the rotating electric generator. Like the rotary turbines, the Stingray can be deployed with a variety of spacings and in staggered arrays. The first model was deployed for testing in Yell Sound, in the Shetland Islands, in 2002–2003 and apparently worked well. The company developing it estimated, based on the initial tests, that Stingrays could produce power at a cost of about ten to twenty cents U.S. per kilowatt hour. Further testing and assessment began in late 2004, funded by a British government grant of US$3.9 million. Ten larger models were scheduled to be installed in the Shetlands between 2005 and 2007 at an estimated cost of US$41 million, resulting in the first Stingray grid-connected tidal farm with an output of five megawatts. But then, in late 2004, the company put the project on hold, at least temporarily, because of insufficient investment monies or other funding to proceed with such a large project.

A more radical concept in tidal power, called tidal sails, is the brainchild of a Norwegian airline pilot named Are Børgesen. He has been an avid racing sailor since childhood, and was a member of his country's national team. During one race, he told me, "I was the tactician on board. We were sailing with the spinnaker up directly against a two-knot current and were just sitting there in the same position," held back by the current even with the wind in their favour. "After a half-hour I thought to myself, what if we turned the boat upside down? The sail would carry us with the current." It would take the boat backward, of course, and with great force. It occurred to Børgesen that, in principle, the energy in the current could be harnessed that way, but not if the boat and sail were allowed to drift freely. "How can we hold a sail against the current?" he asked himself, because only by resisting the current could physical "work" be extracted from it.

These thoughts evolved into a system, now at the prototype stage, in which a series of large, stiff rectangular "sails" made of lightweight glass-reinforced plastic are attached at their four corners to a long set of continuous undersea cables. The cables are suspended under tension above the sea bottom, and they loop around large pulley-like wheels at each end of the cable run. The sails are spaced in an optimum way along the cables, to avoid interfering with the water flow, so that the set-up somewhat resembles a chairlift on a ski slope, with each chair evenly spaced. Deployed in a channel with swift tidal currents, the sails are swept along with the current, thereby pulling the cables along with them and turning the wheels at the ends. At one end, the rotation of the wheels drives an electrical generator. Because the sails are moving in the same direction as the current, they should do no harm to fish or marine mammals.

But what happens when the moving sails reach the end of the cable run? That is where the concept gets a little tricky. They have to be detached from the cables and stored in a large undersea module — Børgesen calls it a magazine — until the tide turns and the current begins to run along the channel in the opposite direction. Then the sails are re-attached to the cables, one by one at the correct spacing, and are swept back to other end of the system, which also has a large storage magazine. The moving cables turn the wheels and generate power regardless of the current's direction. Anywhere from several hundred to one thousand thin sails would be deployed along a 1,000-metre (0.62-mile) long track in the course of a six-hour tidal period.

Devising the attachment and detachment procedure was something of "a headache," Børgesen admitted. "We've been through several alternative ways of doing it. Now we have a solution that is very simple and elegant, but we cannot give out the details." Although his Tidal Sails company has obtained patent protection for its invention, he is still concerned about infringement in some countries and wants to be the first to offer the technology. "China is a huge market," he said. "Britain is definitely one of the best places for our system, and British Columbia might be even better."

For any long tidal channel, Børgesen envisages placing numerous individual cable runs, each with rotating wheels and magazines for sail storage on the ends and a large undersea electrical generator at one end. They would be oriented to slant across the channel at a thirty-degree angle, because this allows them to develop hydrodynamic lift and generate more power than simply by running directly downstream with the current. (A sailboat also sails faster if it is a bit "off the wind.") "Going diagonally across the current gives double or triple the power, and it places the ends," where the magazines, wheels, and generator have to be attached to the sea bottom "on the shallow sides of the channel."

The spacing between parallel cable runs would be about 200 metres (656 feet), but each full-scale system would be custom designed for the specific channel and its conditions. If large ships used the channel, the system would be deployed at least 20 metres (66 feet) below the surface. There would be a bit of sag along the cables between the end stations, but not much. The buoyancy of the sails can be adjusted, and the tension on the cables can be enough to avoid having them swipe the sea bottom.

There are great advantages in efficiency to his system compared to the competing ones, Børgesen insisted. "We have a huge area exposed to the current — the square footage of all the sails combined — for each expensive generator and gearbox," compared to the much smaller surface area of a tidal turbine's blades. To maximize the output from a long tidal channel might require scores or hundreds of individual turbines, but only dozens of tidal sail tracks. "Each will generate tenfold what a turbine can do." There would only be two footprints on the sea bottom per cable run, one at each end, whereas each tidal turbine has to be fixed or anchored individually to the bottom, even though its output is so much smaller.

"We are now in the process of designing the first pilot project, a 30-metre (98-foot) long system to be deployed at an island on the west coast of Norway," Børgesen said. Next he planned a 200-metre (656-foot) long prototype elsewhere on the Norwegian coast, and in 2008 a full-scale prototype. After that will come "the first significant power production." It's "a radical, disruptive technology," he admitted. "You're in completely new territory." But he has studied the costs of the materials. The generators are available essentially "off the shelf," with no design breakthroughs required. "We have the world's largest producer of galvanized steel cable on side." (Skanrope, a Norwegian company.) So he is confident that his system can produce commercial power with only four knot currents at a cost of between four and six cents U.S. per kilowatt hour, and at that price he thinks there should be a vast worldwide market for his invention. "We'll really park the competition," he boasted.

Tidal Power's Promise and Limitations

Whether it will be in the form of tidal lagoons, rotating turbines, nodding hydroplanes on long arms, large sails slipping along with the current, or some other technology yet to be invented, tidal power is bound to play an important role in our world's future. There is simply too much available energy in the tides for it not to be exploited eventually. Most of the new concepts and proposals have only emerged in the last decade or so, however, and to get the tides to yield their energy for mankind's use will involve exciting but daunting engineering challenges.

Because the action of the tides is spread out over entire oceans, it is hard to imagine that we will ever harness more than a minute fraction of that energy. Moreover, as Tom Foley pointed out when I spoke with him at the Annapolis Royal power station, the inherent problem with tidal power is that it is intermittent. For hours at a time, even the best tidal station will generate no power at all. The timing of the tides can be quite different, however, on shorelines not too distant from each other (on the opposite coasts of Britain, for example, or on opposite sides of Vancouver Island), which could significantly even out the total flow of tidal electricity to a power grid. Energy can also be stored, such as by pumping water uphill to a reservoir (to be released through hydroelectric turbines when needed) or by generating hydrogen via electrolysis and using it when needed in fuel cells.

Still, for any power grid to be efficient and reliable, there will likely always have to be large-capacity sources of steady power, be they hydroelectric dams, thermal plants burning fossil fuels (coal, oil, or gas), or nuclear plants. On the plus side, though, the tides and tidal currents are almost perfectly predictable, whereas other renewable energies are not. The wind does not always blow, for example, and thick clouds can reduce the output of a solar system. In principle, then, tidal power can be better integrated into an electrical grid than other intermittent renewable sources. This clean form of endless energy may not, in itself, solve the world's energy problems, but it can make a very significant contribution as part of a mix of carbon-free energy sources.

Once and Future Tides

A Receding Moon and Drifting Continents

Archaeologist Don Morris looked out at the shoreline of Santa Rosa Island and pointed to areas where he thought undersea exploration might turn up traces of human habitation dating to the very end of the Ice Age. We were sitting at a simple grass airstrip on the eastern side of the dry, rugged 25,000-hectare (62,000-acre) island, waiting for the small chartered plane that would take us back across the Santa Barbara Channel to Camarillo, California. Morris was about to retire after a career with the U.S. National Park Service, including a decade and a half as the archaeologist responsible for Channel Islands National Park, of which Santa Rosa is a part. He had just taken me in a four-wheel drive Park Service vehicle to see an excavated site at Arlington Canyon, one of the most important such places in North America.

The bones of the so-called Arlington Woman had been found there in 1959, and decades later advanced radiocarbon dating showed them to be about thirteen thousand years old. Within the margins of error, this made Arlington Woman equal in age to the oldest well-dated human remains ever found in the Americas. She was also contemporaneous with what were long considered to be the oldest artifacts on the North American mainland, Clovis spear points, which had been used to hunt bison and mammoths across much of the American West.

What was remarkable was that she had died on an island that is part of an island group separated from the California mainland by about nine or ten kilometres (around six miles) of water with swift tidal currents. This implies that late Ice Age people must have had a fairly advanced boating technology. At the time, I was working on a book, *Lost World*, about the search for traces of ancient peoples and their migrations along the North Pacific coast, which is why I was there with Morris that day in 2000. And it turned out that the story of the Channel Islands and the local tides is more complex than it would appear at first glance.

Today's tidal range along that part of the California coast is about two metres (six and a half feet). In the Santa Barbara Channel, tidal currents run at speeds up to around two knots in a chaotic, swirling pattern that changes and reverses direction with each ebb and flood. But at the time of Arlington Woman, it was a very different place. At the peak of the last glaciation, some eighteen to twenty thousand years ago, much of the northern hemisphere was buried under great ice sheets up to several kilometres (one mile or more) thick. (In North America, they extended south to Puget Sound on the Pacific coast, and to New York City on the Atlantic.) So much water was locked up in ice that worldwide sea level was as much as 120 metres (nearly 400 feet) lower than it is today. The Bering Strait had disappeared, and Asia and North America were linked by a broad land bridge. The Channel Islands, where we sat, were also linked to each other (but not to the mainland) when sea level was low. They formed a single, much larger ancient island, ten times the size of today's Santa Rosa, that archaeologists call Santarosae. "And by thirteen thousand years ago," Morris told me, "they still would have been connected."

The channel between Santarosae and the California mainland was narrower than it is now, and much shallower. Today, there are passes between the major Channel Islands where tidal currents flow and some of their energy is dissipated. But when there was only one very large island, the tidal currents would have been restricted to the one main channel. The narrower (and shallower) channel "implies swifter currents," Morris said. "In a way, it was like a big, broad, nasty river. It was still five miles [eight kilometres] wide with a strong current." So people would have needed seaworthy boats or rafts to get across to Santarosae from mainland California.

Sea Level Change Causes Intertidal Zones to Migrate

Stronger tidal currents in places like the Santa Barbara Channel were only one of the many changes in the tides resulting from the last glaciation and its effect on sea levels. With world sea level so much lower on average, all the world's shorelines (and, of course, their intertidal zones) would have been in very different places from what we observe today.

To understand this, just look at the depth measurements on a nautical chart, and then peel away 120 metres (395 feet) of water. San Francisco Bay would be entirely empty of sea water. So would Chesapeake Bay, where the average depth today is only 7 to 8 metres (about 25 feet). Then imagine a time-lapse video of the world's shorelines since the Ice Age. As the great ice sheets melted, all that water returned to the oceans and sea level rose. The process began slowly more than fifteen thousand years ago, and the Big Melt picked up speed around thirteen thousand years ago, roughly the time that the sea encroached on the Bering Strait, once more separating North America from Asia. By around nine thousand years ago, most of the ice had melted. (By convention, geologists place the end of the Ice Age at ten thousand years ago.) Sea level continued to rise, but more slowly. Average world sea level nearly stabilized by around seven thousand years ago, although there were significant local changes after that time that had great impact on the tides (and the location of intertidal zones) in particular regions.

Watching that time-lapse photography for any region, we would see large embayments, such as San Francisco Bay, gradually fill up with water. As they do so, we can imagine the tides from the open ocean surging in and out of those bays, creating new intertidal zones. Within those zones, all the familiar organisms would establish themselves: the clams and oysters, the eelgrass beds, the seaweeds and barnacles on intertidal rocks. In many parts of the tropics, mangroves would move in and establish themselves.

But how stable would these post–Ice Age intertidal zones be? Not very. After all, over the span of thousands of years, average sea level was rising rapidly. If we do a rough, back-of-the-envelope calculation and assume an average tidal range of 3 metres (10 feet) — in some places it would be larger, in others smaller — a rise of 120 metres (400 feet) means that there would be about forty "steps" or increments (of 3 metres or 10 feet each) between the lowest sea level at the peak of the Ice Age and today's sea level. And most of those steps would have occurred in about four thousand years (between thirteen thousand and nine thousand years ago). In other words, on average an intertidal zone would exist for only about one hundred years. During that one century, sea level would rise enough that many of the plants and animals living at any particular level or elevation within an intertidal zone would find their situations untenable. Many barnacles, for example, only thrive where they are periodically exposed by the tides to the air. Some seaweeds are also highly sensitive to their elevation above extreme low tide. As sea level rose, they would have to move, or die. Individual barnacles and seaweeds do not physically move, of course, but they can spread and establish new colonies nearby. Each century, all those colonies of intertidal organisms would be forced to migrate to maintain their relative places in the ever-changing kaleidoscope of tidal shorelines, in the web of intertidal life.

This average worldwide sea level rise is only one part of the picture. In many places there would be equally dramatic local effects that would affect tidal shorelines in particular regions. For example, on the British Columbia coast at the peak of the Ice Age the great ice sheets covered nearly the entire shoreline out to the continental shelf with ice up to one or two kilometres (about one mile) thick. The enormous weight of the ice pressed down on the somewhat flexible crust and mantle of the Earth, depressing it as much as 100 to 200 metres (328 to 656 feet). As the ice melted back, the weight came off while the land was still (temporarily) depressed. At Courtenay on Vancouver Island, when the sea returned as the ice retreated, sea level was initially 150 metres (492 feet) *above* today's level. There are places high in the forests of the coastal hills with fossil beaches, where sand, shells, and other remains testify to post–Ice Age intertidal zones. Then, as deglaciation progressed, the land slowly rebounded. Along much of B.C.'s inner coast, this means that sea level *fell* quite drastically, rather than rising. In such regions, intertidal zones would have been forced to migrate seaward, rather than landward, as in many other parts of the world.

The greatest (and fastest) changes in sea level had occurred by around seven thousand years ago. However, worldwide sea level has continued to rise since then, albeit much more slowly. In addition, gradual, but still significant, isostatic rebound (uplift) continues today in some regions, and subsidence of land in others. In Holland, for example, the land is subsiding in a complex and delayed response to the retreat of Ice Age glaciers from Britain and Scandinavia. That is why the sea has been encroaching on the Dutch coast throughout modern history, requiring a never-ending battle to build dikes and protect the land. Over those centuries, the intertidal zones would have migrated as well.

I once observed a similar phenomenon, but with a different cause, while sailing with friends along the north coast of British Columbia, among some sheltered islands not far south of Prince Rupert. We anchored for the evening and rowed in our dinghy along the shore. The tide seemed to be exceptionally high, but then we noticed something strange. The sea was actually lapping at the base of dead standing trees on the gently sloping shoreline. Looking closer, we realized that all the trees from the high tide line back to at least twelve to fifteen metres (forty to fifty feet) were gray and bare of foliage, apparently killed off by the salt water. But they had not yet fallen over. We knew that this forest could not have become established in the first place if salt water had always soaked the soil in which they grew. The sea must have risen around one metre (three feet) within the lifetime (perhaps one hundred years) of these very modestly sized trees.

The Pacific Ocean (like all the world's oceans) *is* in fact rising (mainly due to global warming). However, average ocean levels have risen much less than a metre during the past century. Rather, we were seeing a local effect: the land on B.C.'s north coast has been subsiding. In this case, geologists have shown that the process is mainly due to the movement of tectonic plates pinching and prodding and reconfiguring the entire northwest coast, dragging the land down in some places, pushing it up in others. Similar processes occur all over the world. The drifting plate on which India sits, for example, collided with the continent of Asia about 50 million years ago, pushing up the Himalayas. This uplift is still occurring today. Mount Everest is getting taller. Where the land is subsiding, as on B.C.'s north coast, the intertidal zone is encroaching farther onto the land, and inevitably the shellfish beds, barnacles, seaweeds, and other organisms will be migrating as well.

The rise or fall of local sea level can also cause another change to tidal shores that is quite different in nature. As I discovered on my visit to the Bay of Fundy, and from scientists who have studied its tidal history, it is the close resonance between the natural period of the bay's waters and the Atlantic tides that generates the world's largest tides. This natural period is highly sensitive to changes in the depth and length of the Bay (which have not remained fixed during the many millennia since the Ice Age) and to changes in the depth and tides of the waters outside the Bay.

The Bay of Fundy does not open directly onto the deep Atlantic. Just outside its mouth is a very large bight (a sweeping indentation on the North American shoreline) called the Gulf of Maine. It extends from Cape Cod to Nova Scotia and has banks, sills, and other shallow areas that affect the propagation of tides from the deep, open Atlantic into the coastal waters. During the last glaciation, the seafloor of the Gulf of Maine was pressed down by glacial ice, and since then it has gradually rebounded while world sea level has risen as well. Over the course of the deglaciation and the millennia following it, tides in the Gulf of Maine only gradually reached the range that oceanographers observe today. A mathematical model developed by David Greenberg of the Bedford Institute of Oceanography in Dartmouth, Nova Scotia, indicates that as of seven thousand years ago, the tidal range in the Gulf of Maine was still only 20 to 50 percent as large as it is today. But it was expanding. By four thousand years ago, the tides would have reached 80 percent, and by about twenty-five hundred years ago, they would have attained the same range as today, a mean of about three metres (ten feet). At that time, local sea level was still around seven metres (twenty-three feet) lower than today, and steadily rising. In the past century, sea level at Halifax, which is on the open Atlantic, not on the Bay of Fundy, has risen about thirty-five centimetres (fourteen inches).

Within the bay itself, studies of the ancient tidal range show that the kind of resonance we see today did not occur until conditions were just right. By taking sediment cores in mud flats and salt marshes, and by radiocarbon dating those cores, scientists such as geologist John Shaw (also of the Bedford Institute of Oceanography) have reconstructed the history of low tide and high tide elevations in the Bay. Deeply buried oyster beds are the main indicator of extreme low tides, while salt marshes show where the highest high tides reached at various times in the past. By around seven thousand years ago, most of the post-glacial worldwide sea level rise had taken place and the land in the region had largely rebounded. (In some places it subsequently subsided, in a secondary response to the earlier uplift.) However, the shallow offshore Georges Bank effectively prevented the ocean tides from creating large tides in the Bay. The tidal range within the Bay was only about two metres (six and a half feet). Sea level just outside the Bay continued to rise slowly (ten to fifteen centimetres, or four to six inches, per century) as the combined result of the melting of what was left of the ice sheets and the subsidence of coastal lands. This was enough to submerge Georges Bank and allow the tides to propagate into the bay.

Around four thousand years ago, the bay's waters approached resonance with the tides in the Gulf of Maine and the open Atlantic. Until then, as sea level rose, the intertidal zones (with their modest but slowly expanding tidal range) would have gradually migrated landward (and higher onto shore) toward their current locations. But with a much smaller tidal range than today, the horizontal extent of the intertidal zones would have been much less as well. There would not have been the extremely wide mud flats that we see today.

Then, with Georges Bank submerged and the depth of the bay's waters slowly increasing, the depth and length of the bay became just right for the bay's natural period of oscillation to match that of the Gulf of Maine and the open Atlantic. Resonance kicked in, which greatly amplified the tides. The tidal range expanded rapidly over the course of only one millennium. By three thousand years ago, the mean tidal range had reached about eight metres (twenty-six feet), and peak spring high tides would have been larger.

Rising high water levels forced the local Micmac Indians to abandon long-established settlements they had on the shores of Minas Basin and Chignecto Bay, the two areas where today's highest tides are observed. The Micmacs have a traditional story that may represent a dim memory of this rapid expansion in the tidal range. In one account, "According to the legend of the Micmacs the god Glooscap became angry because a beaver built a lodge in his fishing waters. He threw rocks at the lodge which now form the Five Islands [on the northern shore of the Minas Basin]. A great flood resulted as the lodge was destroyed."

Over the past three thousand years, average sea level in the bay has continued its slow rise. The Earth's crust in the region has subsided further while worldwide sea level continues to creep upward. But the level of salt marsh sediment (representing extreme high spring tides) has risen far more, by a full 7.5 metres (24.5 feet). Near-perfect resonance, therefore, has resulted in the maximum tidal range of over 16 metres (about 54 feet) that can be observed today.

Giant Ancient Tides and the Earth's Rotation

The ten-thousand-year period of changing sea level and migrating tidal shorelines since the end of the Ice Age may sound like a long time. But the Earth has existed for more than four billion years, and life in the Earth's oceans for over three billion. Seen against this scale of deep geological time, ten thousand years is the mere blink of an eye. Not surprisingly, changes in tides and tidal patterns over that vast expanse of time have been enormous. The record of those changes is written in limestones, corals, and seashells. And the story they tell is of nothing less than a great transformation in the Earth-Moon system, one that will continue to evolve for billions of years into the future.

In 1695, Edmund Halley, the British astronomer who sponsored the publication of Newton's *Principia*, was the first to notice a strange phenomenon. Looking back at ancient Middle Eastern records of solar eclipses, Halley saw unexpected discrepancies and realized that during his lifetime the Moon appeared to be moving more rapidly across the sky (in other words, accelerating relative to the distant background stars) than it had several thousand years earlier. In 1754, the German philosopher Immanuel Kant indulged in a bit of entirely speculative reasoning. He wrote a (largely ignored) newspaper article suggesting that the friction of the ocean tides, acting on the seabed and shorelines, must be slowing down the rotation of the Earth, which would make it *look* as though the Moon were orbiting slightly faster. In 1848, the German physicist Julius Robert Mayer made a link between the two ideas, suggesting that a more slowly rotating Earth must also affect the Moon and its orbit. As the rotation of the Earth slowed down slightly over thousands and millions of years (due to what is today called tidal braking), the Moon would have to orbit farther and farther from the Earth.

This is because (according to Newtonian physics) the total energy of a physical system is always conserved. For example, a bullet weighs very little but moves extremely fast, so it has considerable momentum. (Momentum is simply mass multiplied by velocity.) When the bullet hits a heavy but moveable target, that target will be jolted, or accelerated backward, even if only very slightly. In other words, the momentum of the bullet has been transferred to the target, and total momentum has been conserved. In the case of rotational systems, such as a spinning flywheel, it is total *angular* momentum (the mass of the flywheel multiplied by the rate of rotation) that must be conserved. But what if the distribution of mass in the rotating system changes? This is what happens when a spinning figure skater pulls her arms in close to her body. She spins faster. When she puts her arms out again, she slows down and usually comes to a stop.

The Earth and Moon constitute just such a rotating system, because they are linked by gravity. The combined system rotates around their common centre of gravity, or barycentre. But the friction of moving water in the Earth's oceans, caused by such things as the flow of tidal currents across the seabed and against the shores of the continents, acts as a very slight brake on the Earth's rotation. Year by year, century by century, the Earth rotates a little bit more slowly. And, since energy is being neither lost nor gained, this affects the gravitationally linked Moon as well. To conserve the total angular momentum of the Earth-Moon system, the Moon has to spiral out in an increasingly distant orbit.

Recall the analogy between a cheerleader's baton and the Earth-Moon system rotating around its barycentre. The large and massive Earth is on one end of the baton and the smaller and much less massive Moon is out on the other end. If the baton were twirled in the vacuum and weightlessness of space, it would spin forever, and angular momentum would be conserved. Now, if the rate of rotation of the very massive Earth slows down for any reason (such as from tidal friction), in order to conserve the system's total angular momentum, the angular momentum of the Moon has to increase. In other words, the Moon has to move farther away from the Earth. As a result, each orbit of the Moon around the

Earth — each lunar month — becomes a little bit longer as well. This is what has been happening over the past few billion years.

Just how quickly is the Earth's rotation slowing down and the Moon's distance from the Earth increasing? And how has this affected tides on Earth? There is disagreement about this among scientists, and some sizeable discrepancies in the estimates derived from different sciences and types of evidence. But certain things are known with great confidence. For example, the Apollo 12 astronauts placed a reflector on the surface of the Moon in 1969. Since then, by bouncing lasers off it, scientists have determined that the Moon is currently moving away at a rate of 3.8 centimetres (about 1.5 inches) per year.

As for measuring the Earth's rate of rotation in earlier times, the first evidence showing that it used to rotate faster came from growth rings in fossil coral discovered by American geologist and paleontologist John Wells in 1962. While studying living corals on Australia's Great Barrier Reef, he noticed that they had diurnal habits. They were active in daytime and more dormant at night. Then he looked at fossil rugose corals dating to the Devonian period (about 415 to 360 million years ago) found in rocks in upstate New York, and saw that the corals had both large ridges and many more very fine ones. He interpreted the large ridges as representing annual growth lines and the fine ones as daily growth lines. Counting the fine ridges, he determined that there had been about 400 days in each year during the Devonian, rather than the 365 days we have now. Later study of corals that lived during the Pennsylvanian epoch (325 to 285 million years ago) showed a tally of growth rings that fell somewhere between these two figures. Wells's research provided the first independent evidence (separate from theoretical geophysical calculations and speculation) for the slowing of the Earth's rotation. Measurements today show that the Earth's rotation is currently slowing at a rate such that every century, the earthly day becomes about .002 seconds longer.

These changes in the Earth's rotation and in the distance between Earth and Moon may sound insignificant, but over hundreds of millions of years they add up, and they imply major changes in the timing and amplitude of tides on the ancient Earth. When the Moon was closer to the Earth, the tides would have been larger, and a shorter day back then had to entail shorter tidal periods as well.

Precisely what the tides on the very early Earth would have been like is a matter of some controversy, largely because scientists disagree about the rate at which the Earth's rotation has been retarded over long geological periods and how fast the Moon has been spiraling away from the Earth. Some think the rate has held fairly steady over hundreds of millions of years; others see evidence that the rate has increased markedly over time. Depending on who is right, tides on the ancient Earth might have been absolutely colossal, or perhaps "merely" quite a bit larger than they are today.

Aside from coral growth rings, there are other natural processes that have left long-term records of the Earth's changing rate of rotation. Among the many kinds of sedimentary rock known to geologists are tidal rhythmites. These are sandstones, siltstones, or mudstones that formed on the seabed in very sheltered estuaries or tidal deltas hundreds of millions of years ago. With each flood and ebb of the tide, currents deposited extremely thin layers of sediment (laminations, or laminae) over long periods of time. Some laminae in the rhythmites are thicker than others, and the differences relate to the tidal range and speed of tidal currents at that location, day by day. In the late 1990s, Australian geologist George E. Williams analyzed 620-million-year-old (Precambrian) rhythmites from southern Australia and determined from them the length of the Earth's day and the distance between the Earth and Moon. At that time, he concluded, the Earth took about 21.9 hours to rotate on its axis (instead of 24), and the Moon's distance from Earth was about 96.5 percent of its current distance. This implies that since that time, the mean rate (averaged over the entire period) at which the Moon has been receding is 2.17 centimetres (0.85 inches) per year, or just over half of the present rate of recession. In other words, the Moon's rate of recession has increased considerably. Williams also looked at some much more ancient (2.45-billion-year-old) deposits in western Australia, for which he had somewhat less confidence that they reflected tides and tidal periods. These suggested the Moon was about 90 percent of its present distance from Earth, and implied a mean rate of recession of 1.24 centimetres (0.49 inches) per year during the time between 2.45 billion and 620 million years ago. Since gravitational attraction between two celestial bodies varies as the inverse cube of their distance apart, a 90 percent distance implies tide-generating forces that were about 37 percent stronger 2.45

billion years ago than they are today. Tidal ranges would have been larger than today, but not drastically so.

Other studies, however, have come up with results suggesting that ancient tides were far larger than that. In the mid-1990s, C.P Sonnet of the University of Arizona and Erik Kvale of the Indiana Geological Society, along with several colleagues, studied rhythmites from Indiana, Utah, Alabama, and southern Australia that were about 900 million years old. They concluded that each earthly day at that time was about eighteen hours long, which meant the Earth's rotation has been slowing down at a greater rate than George E. Williams's Australian data showed. Implied also would be that the Moon was considerably closer to the Earth nearly 1 billion years ago, and even closer during the preceding billions of years. The tides would have been proportionally far larger as well.

Aside from fossil coral, there is other biological evidence for the rate at which the ancient Earth rotated. The chambered nautilus, a cephalopod related to octopuses and squids, has a shell with separate chambers that the animal periodically seals off by secreting a wall as it grows. Between each pair of walls are many fine laminations. Today's nautiluses produce about thirty laminations between their walls, which suggests a relationship with the (roughly) thirty-day lunar month. Each lamination represents one day. When paleontologists look at ancient fossil nautiluses, however, the farther back in time they go, the fewer the laminations between the walls, and this decrease seems to occur in a consistent, almost linear, fashion. The most recent known fossil nautiluses have about twenty-five laminations between each of their walls, but 420 million years ago there were on average only about nine or ten laminations. This would imply a lunar month of only nine or ten days, with the Moon revolving in an orbit only about 40 percent of its present distance from the Earth, and drastically more powerful tide-generating forces. The Moon would have looked like a gigantic orb dominating the night sky, and much larger in apparent diameter than the Sun. (Today, the Sun and Moon look nearly the same size, which is why the Moon almost exactly blocks the Sun during a solar eclipse.) Since gravity varies as the inverse cube of the distance between celestial bodies, the ancient tide-generating forces would have been more than fifteen times as strong as today. Tides on Earth would have been absolutely colossal.

Whatever the exact rate of lunar recession, the ancient Earth had a moon that was much closer than it is today, and tides, on average, would have been anywhere from significantly larger to truly enormous. The consequences for life on Earth might have been considerable. In fact, some scientists have speculated that without the existence of the Moon and large tides, life might not have emerged on Earth at all.

Today's dominant theory of the Moon's origins is that it was formed from a portion of the crust of the very early (and still soft or molten) Earth. That material was propelled out into space by the oblique collision of another large body (a huge asteroid, comet, or other planetesimal) with the Earth. This event has been called the "big whack" or "big splash," and it resulted in a moon that initially revolved around the Earth in a much closer orbit than it does today. Astronomers believe that the Earth first formed about 4.5 billion years ago, at which time it is believed that the Earth rotated in about six hours. Then came the "big whack," followed by solidification of the Earth's crust. Oceans must have formed next (possibly from water brought to Earth by comets), and life appeared around 3.6 to 3.9 billion years ago, all of which is quite fast by astronomical standards.

Although some scientists think life could have been introduced to Earth from space, most biologists believe that it originated in the oceans. Many reckon that this first occurred in huge, warm intertidal pools that would have been created on the shores of the ancient Earth by gigantic tides. As early as 1871, Charles Darwin mused in a letter to a friend that life might have first formed in "some warm little pond with all sorts of ammonia and phosphoric salts." Support for this viewpoint came with the famous Miller-Urey laboratory experiment of 1953, when a warm mixture of methane, ammonia, hydrogen, and water was subjected to an electrical discharge. The result was organic molecules, sugars, and amino acids, the building blocks of life. On the ancient Earth, the mixture of chemicals in warm sea water (and struck by lightning or exposed to energetic radiation from space) would have constituted what scientists call the "primordial soup."

Some scientists have further speculated that large tidal pools on the shores of ancient seas could have provided the ideal conditions for the generation of life. Such pools would have repeatedly filled up with warm water, dried out under a powerful sun (at a time when there was little or no ozone to filter the radiation), and then filled up again on the next tide. This kind of hydration-dehydration cycle has been found in the laboratory to produce larger and more complex organic molecules, which biochemists call polymerization. This means that strong tides, delivering larger volumes of water for the hydration-dehydration cycle, could have facilitated the development of the self-replicating molecules in the primordial soup. As astronomer C.R. Benn of Britain's Isaac Newton Group summarizes this ongoing speculation on the origins of life, "One common theme to emerge is the importance of concentrating the soup to encourage polymerisation, e.g. in tidal pools which repeatedly dry out under the sun."

There are even astronomers who think that when the material that makes up the Moon first flew out from the Earth and formed our satellite, it might have orbited initially as much as ten times closer to us than it does today. (Any closer and the Earth's gravity would have torn it apart.) "In this case," says physicist and astronomer Neil Comins of the University of Maine, "the tides on the young Earth were 1,000 times higher than they are today, since tidal forces vary inversely with the cube of the distance. These humungous tides plunged miles inland and withdrew every three hours," because with such a close Moon and the Earth rotating much faster, the day would have been only six hours long. As these gigantic tides "moved over the land, the awesome volumes of water scraped and pounded the primeval rock, removing and pulverizing a considerable amount of it. Every time the tide retreated, it dragged this material back into the ocean. Continually churned up by the water, these chemicals formed the broth in which life probably formed."

Whether tides were responsible for the origins of life on Earth or not, early life thrived initially only in the oceans. (It took 2 to 3 billion years for the atmosphere to reach the balance of oxygen and nitrogen that exists now.) Life did not find its way onto land until around 500 million years ago, a time when, most scientists believe, the tides were at least somewhat larger than they are now. Large intertidal zones may have provided a particularly well-suited transitional zone that helped life — first plants, and later animals — to emerge and colonize the land.

Comins, in his book *What If the Moon Did Not Exist?*, describes how this process might have occurred: "Land animals are descendants of aquatic life. On Earth, the transition to life on land came when amphibians crawled or were deposited on beaches by tides nearly 400 million years ago. These creatures primarily breathed oxygen dissolved in water through their gills. They had limited capacity to breathe the oxygen-rich air on the earth's surface." Why? Because, as Comins goes on to say, "oxygen in high concentrations is toxic, making the evolution of lungs that could limit the amount of oxygen entering an animal's bloodstream an important development. Fortunately for our ancestors on Earth, the ocean tides of 400 million years ago had decreased to little more than they are today." They were "little more," that is, compared to the "humungous" earlier tides that Comins posits. He continues, "Therefore, our ancestors emerged from the oceans during relatively calm high tides, stayed on land a few minutes, and returned to the ocean before the water receded too far for their flipper legs to carry them."

Those tidal shorelines would not have been at all the same as what we know from modern observations. For example, because of the movement of tectonic plates (continental drift), some 225 million years ago all the major continents were bunched together in a super-continent that geologists call Pangaea. Around it in every direction would have stretched a single huge ocean, and it is fair to assume that the action of tides in that ocean would have been quite different from what occurs in the more fragmented ocean basins of today. The North Atlantic Ocean only began to "open up" about 60 million years ago, as North America and Europe drifted apart, and it did not reach its current depth and approximate dimensions for tens of millions of years. Until then, the pattern of tides on the shores of the Atlantic likely would have differed from what we see now.

Nor has the story of tidal evolution reached any kind of final chapter. Let's fast-forward the videotape of the Earth's geology and the gradually changing Earth-Moon system. If tides on the Bay of Fundy did not attain their present resonance with the tides of the Gulf of Maine and the Atlantic until several thousand years ago, what about several thousand years into the future? Assuming that global warming continues, world sea level may eventually rise enough (due to gradually melting ice sheets and the expansion of warmer sea water) to change significantly the depth, length, and shape of the Bay of Fundy, possibly weakening the resonance and reducing the tidal range there. At the same time, other bays and estuaries elsewhere in the world may attain the characteristics required for resonance and extremely large tides. And whatever their tidal ranges, intertidal zones everywhere will continue to migrate.

On the other hand, what if human-caused climate change does not continue over the next few thousand years? In that case, another pattern of global climate variation could still come into play. Astronomers and geologists recognize what is called the Milankovitch cycle, named after the Serbian engineer and geophysicist Milutin Milankovitch, who developed a theory of ice ages during the first half of the twentieth century. Milankovitch argued that the Earth's climate varied over many millennia, mainly due to periodic variations in the strength and seasonal timing of the Sun's radiation (or insolation) as it hits the Earth.

The Earth's movements change gradually over tens of thousands of years. Over a cycle of roughly one hundred thousand years, for example, its orbit around the Sun becomes more or less eccentric, meaning that it is significantly more elliptical (or oblong) in shape at certain times, but very close to circular at others. At its most elliptical, 23 percent more solar radiation reaches the Earth at perihelion (the time of the year when the Earth is closest to the Sun) than at aphelion (when it is farthest away). The tilt of the Earth's axis of rotation also varies (or changes in obliquity) relative to the plane of its orbit around the Sun. This occurs on a cycle of about forty thousand years. When the angle of tilt is at its greatest, the amount of solar radiation reaching each of the Earth's hemispheres varies more over the course of any year than when the angle is less. Finally, the orientation of the Earth's axis of rotation with respect to the Sun at times of aphelion and perihelion changes over a cycle of about twenty thousand years, much as a spinning toy top wobbles and wanders in a circle as the top slows down. This is called the precession of the equinoxes. Milankovitch proposed that these cycles interact to produce a periodic alternation between glaciations and inter-glacial epochs, such as the inter-glacial epoch (called the Holocene) that we live in today, which has already lasted about ten thousand years.

Two or three decades ago, most climate scientists thought that the Holocene climate had reached its warmest about six thousand years ago and that, since then, the Earth has been in a long-term cooling trend that might last about twenty thousand years or more. Some predicted that we would be entering a new glaciation within perhaps ten thousand years, or even sooner. If so, the great ice sheets of the last glaciation would re-establish themselves across the north of Canada and Siberia, and the immense volume of water locked up in them would lead to a drastic drop in sea level. Just as a rise in sea level would affect the resonance of a place like the Bay of Fundy, so would a drop in sea level, reversing the huge expansion in tidal range there that has occurred in the past seven thousand years. Inevitably, falling sea level would have a major impact on the tides and the location of intertidal zones around the world.

But this scenario no longer appears likely. The 1980s and 1990s brought the growing recognition of carbon dioxide buildup in the atmosphere, and climate scientists revisited the simple models that underlay the idea that the Earth was about to get colder relatively soon. Today, based on more sophisticated models that take carbon dioxide into effect and simulate changes in the Greenland and Antarctic ice sheets as well as ocean circulation, many climatologists think that an irreversible greenhouse effect could overwhelm the predicted ups and downs of temperature based on the Milankovitch cycle. This might perpetuate the current warm inter-glacial epoch for tens or hundreds of thousands of years. If they are right, today's ice sheets will melt and shrink, with drastic consequences for the world's shorelines and all life along them. There are still many unknowns, but the collapse of the West Antarctic Ice Sheet alone would probably raise worldwide sea level by some 6 metres (about 20 feet). If all Antarctic and Greenland ice eventually melted, sea level is projected to rise by close to 70 metres (230 feet). Intertidal zones would move far inland from any shores that have existed in recent geological time.

What if we continue fast-forwarding our videotape millions of years into the future? Then plate tectonics and continental drift come increasingly into play. The picture gets a bit bizarre, but let's have fun and speculate a bit. As we have seen, the Atlantic Ocean did not exist until about 60 million years ago, as the continents fringing it diverged. Projected plate movements indicate that the Atlantic will continue to widen over the next 50 million years, resulting in an ocean that might be large enough to encompass several additional amphidromic systems. Meanwhile, the Pacific will shrink in breadth and Africa will collide with

Europe, closing the Mediterranean. Finally, part of coastal California, which is free to drift, will move northward, sliding along the main North American continental landmass to southern Alaska, which could make for much larger tides on California's no-longer-balmy shores. Then, 250 million years from now, most of the continents are expected to be clustered together again, as they once were with ancient Pangaea, to form a new super-continent with a vast ocean surrounding it. What the tides would be like in that ocean is anyone's guess.

And how about billions of years into the future? On that time scale, the evolving Earth-Moon system itself will have the dominant influence on the tides. As the Earth's rotation continues to slow down, the Moon will move ever farther and farther away and take longer to complete each orbit around the Earth. The tug of the Moon on the Earth's oceans will become less and less, so lunar tides will become ever smaller. At some point the Moon's gravitational effect will become weaker than that of the Sun, and solar tides will begin to dominate, thereby changing the nature of the spring-neap cycle.

Astronomers hypothesize that in around 50 billion years the Earth's day will be forty-seven times longer than it is now. The Moon will be orbiting about twice as far from Earth as it does today and will take forty-seven of our current days to revolve around the Earth. This means that only one side of the Earth will ever face toward the Moon (just as today only one side of the Moon ever faces the Earth). They will be gravitationally "locked." With no net earthly rotation relative to the Moon, there will be no ebb and flow of lunar tides, only the much smaller solar ones. (All of this presumes, of course, that the Sun is still burning and the Earth retains a liquid ocean.) And the Earth will be rotating so slowly by then that each rise and fall of the small solar tides will seem to take forever.

For all practical purposes, the tides as we have known them, coped with them, and depended on them throughout human history will cease to exist. These changes will certainly present major hardships for people who try to derive their food from the (much smaller) intertidal zones of the distant future. Fortunately, this scenario will take a while to play out. In the meanwhile, there is still plenty of time to go out and dig some clams.

Epilogue

Ah, digging clams. Even after more than thirty years living on the B.C. coast, it is still one of my favourite pastimes.

On Salt Spring Island, where I live, there are a number of good beaches for harvesting clams. During the months with an "r" in them, I keep an eye on the tide table published in our local weekly newspaper, looking for convenient daytime low tides. (Digging clams at night, by flashlight or kerosene lantern, is more tedious, although I've done it many times.) Then I put on my gumboots, take a galvanized bucket and the four-pronged digging fork that I use in my vegetable garden, and head for the beach. It's a lovely spot, next to a wooded country road. When I stop to rest, I can look out across a wide and quiet channel to sparsely populated neighbouring islands. There are almost always birds patrolling the mud flats nearby — seagulls, ravens, herons — and often seals, sea lions, or river otters cavorting or swimming by offshore.

Where I dig, there are several species of clams, plus oysters and cockles as well. The cockles, I have found, are too rubbery to be worth the taking. The oysters are good but sparse, and if I want a whole meal of them, I know of a better spot. Usually, it is only the clams that I am after. There, too, I am quite choosy, and I like to think that shellfish gatherers throughout human history have usually been equally selective.

Buried in "my" beach there are lots of really big horse clams. Two or three dozen would probably fill my bucket. Although their meat and juice are tasty, the meat tends to be tough, so I avoid them. There is another, somewhat smaller, clam that has a fragile and elongated shell. I think its proper name is the macoma clam. They are abundant in some levels of the intertidal zone, but they are usually full of sandy grit, which is difficult to remove.

What I like to dig are the little neck or manila clams, which have subtle but colourful ring-like or chequered markings on their shells. They are small but have tender meat and very little grit, which makes them perfect for the culinary delight I have in mind for them. As I dig, my mouth begins to water when I think of that old family recipe. It is a simple Italian-style white clam sauce for spaghetti, a dish that takes me back to my childhood.

My "uncle" Leo Charwat (actually my father's first cousin) was a stockbroker who cooked up a clam sauce only once or twice a year. But when he did, it was with ceremony and reverence, a drawn-out procedure that took several hours. He had nary a drop of Italian blood in his veins, but somehow he had acquired a taste for *linguine con vongole*. And it was so richly flavoured, so pungent with the pure taste of fresh, tender, tiny clams that I can still conjure it up from my memory. That's the taste I try to emulate when I make my own clam sauce today.

I begin by scrubbing the clams to rid them of any slime or grit on the outside of the shells. Then I place them in a very large pot with just a little fresh water, put on the cover, and steam them open. The clams exude their own rich broth. Meanwhile, I sautee minced garlic, onions, parsley, and oregano in olive oil and season the mix with ground white pepper. When the steamed clams and broth have cooled enough to handle, I strip out the clams from the shells and swish them around a bit in the clam broth to rid them of any remaining grit. Then I snip off the tough black siphons (the "little necks") and discard them. I strain the clam broth through fine cheesecloth or a paper coffee filter to eliminate the grit and bits of clam shell and add this purified broth to the sautéed garlic and olive oil mix.

I boil down the liquid a bit to concentrate the flavourful sauce. The cooked clams go in last, once the rest is hot and has simmered for a while, and the sauce is ready to serve over pasta — I personally prefer vermicelli, rather than linguine — with a garnish of grated parmesan or romano and fresh parsley. *Basta*, as the Italians say. A simple but sublime meal.

But first I need to harvest the clams. I find them about a hand span deep in the mix of sand, mud, and fine gravel that makes up the intertidal flat. With each thrust of my digging fork, I turn over a big, wet glob of the muck and look for a telltale squirt of water. Then I reach down and feel around in the sand and cold sea water that quickly seeps into the hole, hoping to find the clam. If the digging is good, I may find one or two of these small clams on each probe with the fork. More often, though, I find at least some dead, empty shells, or other kinds of clams that I reject. It usually takes me about an hour to fill my bucket.

By then, the tide is often creeping back in, flooding the holes I've dug and swirling around my gumboots. I heft my bucket and carry it up the beach toward the road. At the back of the beach, above the line of driftwood and dead seaweed, is an eroding embankment where the roots of the roadside trees protrude. And so do fragments of white shell. The entire raised shoreline is a shell midden that runs along this part of Salt Spring's coast as far as I can see. It results from thousands of seasons of ancient shellfish digging by native Coast Salish families, who came here every year to harvest clams and left the shells behind.

I walk past that midden and up to the road, pleased that I have the makings of several wonderful meals. And as I do, I silently acknowledge my link to those First Nations people and to the millions of others around the world who have lived by the tides for countless millennia.

Further Reading

Chapter 1

N.K. Pannikar and T.M. Srinivasan. "The Concept of Tides in Ancient India." *Indian Journal of the History of Science*, Vol. 6, pp. 36-50, 1971.

Margaret Deacon. *Scientists and the Sea, 1650–1900*. New York: Academic Press, 1971. Chapters 1 & 2.

David Edgar Cartwright. *Tides, A Scientific History*. Cambridge: Cambridge University Press, 1999. Chapters 2 & 3.

George Howard Darwin. *The Tides and Kindred Phenomena in the Solar System*. San Francisco: W.H. Freeman and Co., 1962. Chapters 3 & 4.

Chapter 2

Bill Proctor and Yvonne Maximchuk. *Full Moon, Flood Tide: Bill Proctor's Raincoast*. Pender Harbour, B.C.: Harbour Publishing, 2003.

To purchase the CD *Saltchuck Serenade* by Brian Robertson, including "When the Tide Goes Out," go to www.cdbaby.com/cd/brianrobertson.

Chapter 3

Margaret Deacon. *Scientists and the Sea, 1650–1900*. Chapters 3, 4 & 5.

David Edgar Cartwright. *Tides, A Scientific History*. Chapters 4, 5 & 6.

George Howard Darwin. *The Tides and Kindred Phenomena in the Solar System*. Chapter 5.

Chapter 4

Simon Winchester. "In the Eye of the Whirlpool." *Smithsonian Magazine*, Vol. 32, No. 5, (August 2001): 84–94.

B. Gjevik, H. Moe, and A. Ommundsen. "Strong Topographic Enhancement of Tidal Currents: Tales of the Maelstrom." Internal paper. University of Oslo, September 5, 1997.

Chapter 5

Con Desplanque and David J. Mossman. "Tides and their seminal impact on the geology, geography, history and socio-economics of the Bay of Fundy, eastern Canada." *Atlantic Geology*, Vol. 40, No.1 (March 2004): 1–130.

Andre Guilcher. *Coral Reef Geomorphology*. New York: John Wiley & Sons, 1988.

Gerhard Masselink and Michael G. Hughes. *Introduction to Coastal Processes and Geomorphology*. London: Hodder Arnold, 2003. Chapters 6, 7 & 8.

P.S. Rosen. "Boulder barricades in central Labrador." *Journal of Sedimentary Research*, Vol. 49, No. 4 (1979) 1113–1124.

Leonard M. Bahr and William P. Lanier. *The Ecology of Intertidal Oyster Reefs of the South Atlantic Coast: A Community Profile*. Washington D.C.: U.S. Fish and Wildlife Service, Office of Biological Services special publication No. 81/15, May 1981.

Gerhard Masselink and Bruce Hegge. "Morphodynamics of meso- and macrotidal beaches: examples from central Queensland, Australia." *Marine Geology*, No. 129 (1995), 1–23.

Chapter 6

Julie Wakefield. *Halley's Quest: A Selfless Genius and His Troubled Paramour*. Washington, D.C.: Joseph Henry Press, 2005.

Diana Preston and Michael Preston. *A Pirate of Exquisite Mind: Explorer, Naturalist and Buccaneer: The Life of William Dampier.* New York: Walker & Co, 2004.

William Dampier. *Voyages and descriptions in three parts, Part 3. A discourse of trade-winds, breezes, storms, seasons of the year, tides and currents of the torrid zone throughout the world.* London: James Knapton, 1699. Available online at www.canadiana.org/ECO/mtq?doc=34673

William Dampier. *A New Voyage Round the World.* New York: Dover, 1968.

Edmund Halley. "An Account of the Course of the Tides at Tonqueen in a Letter from Mr. Francis Davenport July 15, 1678, with the Theory of Them, at the Barr of Tonqueen, by the Learned Edmund Halley Fellow of the Royal Society." *Philosophical Transactions*, Vol. 14 (1684): 677–688.

David Edgar Cartwright. *Tides, A Scientific History.* Chapter 6.

Chapter 7

Joshua Slocum. *Sailing Alone Around the World.* New York: The Century Company, 1900.

Chapter 8

Michael Sean Reidy. "The Flux and Reflux of Science: The Study of the Tides and the Organization of Early Victorian Science." Ph.D. thesis for University of Minnesota, February 2000. UMI (microform) No. 9957675.

Admiral G.S. Ritchie. *The Admiralty Chart: British Naval Hydrography in the Nineteenth Century.* London: Hollis & Carter, 1967. Chapter 13.

Margaret Deacon. *Scientists and the Sea, 1650–1900.* Chapter 12.

David Edgar Cartwright. *Tides, A Scientific History.* Chapter 9.

Chapter 9

John and Mildred Teal. *Life and Death of the Salt Marsh*. Boston: Atlantic Monthly Press, 1969.

Chapter 10

Paul D. Kormar. *Beach Processes and Sedimentation*. Englewood Cliffs: Prentice-Hall, 1976. Chapter 5.

David Edgar Cartwright. *Tides, A Scientific History*. Chapters 7 & 8.

George Howard Darwin. *The Tides and Kindred Phenomena in the Solar System*. Chapter 11.

David T. Pugh. *Tides, Surges and Mean Sea-Level*. New York: John Wiley & Sons, 1987. Chapters 4 & 5.

Chapter 11

John R. Harper et al. "Final Report, Broughton Archipelago Clam Terrace Survey." Sidney, B.C.: File P95/16, Coastal & Ocean Resources, Inc., October 23, 1995.

Con Desplanque and David J. Mossman, "Tides and their seminal impact on the geology, geography, history and socio-economics of the Bay of Fundy, eastern Canada."

David T. Pugh. *Tides, Surges and Mean Sea-Level*. Chapter 6.

Chapter 12

Charles D. Keeling and Timothy P. Whorf. "Possible forcing of global temperature by the oceanic tides." Proceedings of the National Academy of Sciences, Vol. 94 (August 1997): 8321–8328.

Charles D. Keeling and Timothy P. Whorf. "The 1,800-year oceanic tidal cycle: A possible cause of rapid climate change." Proceedings of the National Academy of Sciences, Vol. 97, No. 8 (April 11, 2000): 3814–3819.

Walter Munk, Matthew Dzieciuch, and Steven Jayne. "Millennial Climate Variability: Is There a Tidal Connection?" *Journal of Climate*, Volume 15 (February 15, 2002): 370–385; and brief comment by Keeling and Whorf, 446.

Sir Robert Chapman. "Tides of Australia." *Year Book Australia 1938*, online at www.abs.gov.au/ausstats/abs@.nsf/94713ad445ff1425ca256 82000192af2/47edb7b37eb8dd76ca256a09001647aa!OpenDocument.

David Edgar Cartwright. *Tides, a Scientific History*. Chapter 9.

Chapter 13

Nigel Holloway. "The Power of the Moon." *Forbes Global* (July 21, 2003): 42–43.

Orkney Renewable Energy Forum. "Tidal Stream Power for Orkney?" Online report at www.oref.co.uk/tidalres.htm.

U.K. House of Commons Select Committee on Science and Technology, Seventh Report. "Wave and Tidal Energy." May 8, 2001. Online report at www.publications.parliament.uk/pa/cm200001/cmselect/cmsctech/291/29102.htm.

Roger H. Charlier. "A 'sleeper' awakes: tidal current power." *Renewable and Sustainable Energy Reviews*, Vol. 7 (2003): 515–529.

The Bay of Fundy Tidal Power Review Board. *The Tides of Fundy: Renewable Energy Resources in the Maritimes*, January 1977.

Adam Westwood. "Ocean Power, Wave and Tidal Energy Review." *Refocus*, Vol. 5, Issue 5 (Sept/Oct 2004): 50–55.

Chapter 14

J.A. Church and J.M. Gregory. "Changes in Sea Level," Report No. 11 of Climate Change 2001: The Scientific Basis, Working Group I. UN Intergovernmental Panel on Climate Change.

J. Shaw and J. Ceman. "Salt marsh aggradation in response to late-Holocene sea-level rise at Amherst Point, Nova Scotia, Canada." *The Holocene*, Vol. 9, No. 4, 1 (July 1999): 439–451.

Prof. Carl L. Amos, "Lecture 23: The evolution of estuaries: the Bay of Fundy." United Kingdom: Department of Oceanography, University of Southampton, n.d.

Peter Brosche. "Understanding tidal friction: A history of science in a nutshell." *Science Tribune* (December 1998). Available at www.tribunes.com/tribune/art98/bros.htm.

C. R. Benn. "The moon and the origin of life." *Earth, Moon and Planets* 85/86, No. 61: 2001.

George E. Williams. "Geological constraints on the Precambrian history of Earth's rotation and the Moon's orbit." *Review of Geophysics*, Vol. 38, Issue 1, 2000): 37–60.

C.P. Sonett, E.P. Kvale et al. "Late Proterozoic and Paleozoic Tides, Retreat of the Moon, and Rotation of the Earth." *Science*, New Series, Vol. 273, No. 5271 (July 5, 1996): 100–104.

A. Berger and M.F. Loutre. "Climate: An Exceptionally Long Interglacial Ahead?" *Science*, Vol. 297, no. 5585 (August 23, 2002): 1287-1288.

E.C. Pielou. *After the Ice Age: The Return of Life to Glaciated North America*. Chicago: University of Chicago Press, 1991.

David T. Pugh. *Tides, Surges and Mean Sea-Level*. Chapter 9.

MEMBER OF SCABRINI GROUP

Québec, Canada
2007